Magnetic Fields Near and Far
Galactic and Extragalactic Single-Dish Radio Observations of the Zeeman Effect

Timothy Robishaw

DISSERTATION.COM

Boca Raton

Magnetic Fields Near and Far:
Galactic and Extragalactic Single-Dish Radio Observations of the Zeeman Effect

Dissertation.com
Boca Raton, Florida
USA • 2008

ISBN-10: 1-59942-684-6
ISBN-13: 978-1-59942-684-6

Magnetic Fields Near and Far:
Galactic and Extragalactic Single-Dish Radio Observations of the Zeeman Effect

by

Timothy Robishaw

B.A. (University of California, Berkeley) 1998
M.A. (University of California, Berkeley) 2002

A dissertation submitted in partial satisfaction of the
requirements for the degree of
Doctor of Philosophy

in

Astrophysics

in the

GRADUATE DIVISION
of the
UNIVERSITY OF CALIFORNIA, BERKELEY

Committee in charge:
Professor Carl Heiles, Chair
Professor James R. Graham
Professor William L. Holzapfel

Spring 2008

Magnetic Fields Near and Far:

Galactic and Extragalactic Single-Dish Radio Observations of the Zeeman Effect

Abstract

Magnetic Fields Near and Far:

Galactic and Extragalactic Single-Dish Radio Observations of the Zeeman Effect

by

Timothy Robishaw

Doctor of Philosophy in Astrophysics

University of California, Berkeley

Professor Carl Heiles, Chair

According to astrophysical theory, magnetic fields should play an important role in the structure and dynamics of the interstellar medium. While astronomical observations confirm this directly, the observational record is sparse. This is because magnetic fields can only be measured via polarimetric methods, and most of these methods can only provide an indirect inference of the magnetic field strength. The Zeeman effect, however, is the only method by which in situ measurements of astrophysical magnetic fields can be made.

The spectral signature of Zeeman splitting is imprinted in the circular polarization spectrum of radiation received from an astronomical source. In order to make a reliable detection at radio frequencies, one must employ careful calibrations and account for instrumental effects. We begin this dissertation by covering the fundamentals of radio spectropolarimetry. We then offer historical details regarding the Zeeman effect and its use in single-dish radio observations. We present an outline of how one accurately measures the Zeeman effect using large single-dish radio telescopes. We follow this with results from an assessment of the polarization properties of the 100 m Green Bank Telescope (GBT).

We then present magnetic field detections made via the Zeeman effect from the Galactic scale to cosmological distances. We begin with GBT observations of 21 cm emission toward the Taurus Molecular Cloud (TMC) complex. Recent observations have suggested that fields stronger than 20 μG are located at the distance of the TMC. Our Zeeman observations rule out fields of this strength, but do show a clear $\sim+5$ μG detection from H I emission at the velocity of the TMC. More surprisingly, we have discovered multiple detections of a line-of-sight magnetic field of

2

strength $\sim$+40 μG in a filament near -50 km s^{-1}. We then present a windfall of detections of milligauss-strength magnetic fields in starburst galaxies. Detected by means of Zeeman splitting of 1667 MHz hydroxyl megamaser emission, these Arecibo and GBT results represent the first extragalactic Zeeman measurements to probe the field inside an external galaxy. Finally, we climb the cosmological distance ladder, and present a dramatic GBT detection of a magnetic field in a damped Lyα absorber at a redshift of 0.692. We discuss possible scenarios for the creation of an 84 μG field at a look-back time of 6.4 Gyr.

Professor Carl Heiles
Dissertation Committee Chair

To my parents and brother,
who supported me while I avoided the real world.

In memoriam   —   *Lillian A. Cashman 1911–2008*
Richard C. Spillane 1946–1997

ii

Contents

vi

List of Figures

List of Tables

x

Acknowledgments

You might be smart up in space, but here on Earth: no common sense.

MARY ROBISHAW

My residency in Campbell Hall began in 1996, when I took my first astronomy lab course with Carl Heiles. During our first observing session, Carl made us point our theodolites at the North Star. However, we had a great deal of trouble keeping the star in the field of view, until our TA pointed out that the star was not, in fact, Polaris. Carl explained that while this was an *optical* astronomy course, he had good reason to not know which star is at the pole: because he was a *radio* astronomer, and he had been educated at *Princeton*. We had so much fun in that class that a group of us took Carl's radio astronomy lab course the next year. We worked harder than was reasonable for a two unit course, but it was only because of the challenge: when you finished a physics problem set, you dropped it in a box; when you finished one of Carl's do-it-yourself projects, you had looked out into a universe that was invisible to the naked eye. I was hooked. I then worked for a few years with Leo Blitz, who taught me a great deal about how exciting real astrophysical research could be. If it weren't for Leo, I wouldn't have chosen astronomy as a career path, and I'm extremely grateful to him for his support and guidance. I then began the work that became this dissertation back under Carl's wing.

Carl provided me with a unique opportunity to do extremely exciting research using the two biggest telescopes on Earth; I have found this work extremely rewarding. Writing, unfortunately, and to Carl's certain dismay, has not come so easily. I was recently comforted to read that the great historian Barbara Tuchman also found that, "Research is endlessly seductive; writing is hard work." I'm hoping it gets easier; at the very least, this thesis was good practice.

From the start, Carl has treated me like a colleague rather than a student, even when I ask questions that clearly expose my lack of understanding of some fundamental astronomical concepts. Every day I'm amazed by how much he can accomplish and how well he understands so many aspects of observational radio astronomy. Carl has taught me how to be an expert observer, and in this age of email-it-in observing, I feel extremely lucky to have had so much hands-on observing experience. The most important thing that I've learned from Carl is the desire to make sure a job gets done right. He's a great teacher, a great astronomer, and an even greater pal. Above all, I can't thank Carl enough for his understanding and support when my family went through a severely difficult period.

I got to know two of my favorite people, Tom Troland and Josh Simon, via long, long observing runs at Green Bank and Arecibo, respectively. Tom is an incredibly meticulous observer, a source of good advice, a great maker of popcorn, and a great guy. I learned everything I know about booze from Tom. Josh amazes me with his fearlessness in tackling observational projects all over the spectrum, and I hope I get to work with him again. He's alright for a Stanford fan.

A number of folks in Campbell Hall made my infrequent daylight operating hours really enjoyable: Mike Cooper, Snežana Stanimirović, Alberto Bolatto, Karin Sandstrom, John Johnson, and Marijke Haverkorn. Thanks to Jon Swift for having the right attitude about this astronomy gig, and to Danny Jones for looking out for me: he's alright for a Yankees fan. And the biggest thanks go to Mike Fitzgerald and Doug Finkbeiner: whenever I've needed to know something, about anything, at any strange hour, these guys have always provided a timely answer. I'm lucky to have met some new friends when astronomy took me away from Berkeley, and they are all extremely reliable folks: Kayhan Gultekin, Neil Nagar, Jeremy Darling, Crystal Brogan, Dick Crutcher, Bryan Gaensler, Loris Magnani, Peter Teuben, Dap Hartmann, and Tom Bania.

Now and then, my real-world friends made sure life wasn't all work. Thanks to: Kevin Cuff for the #55s and stickball, Pat Mahoney for the hegna, Sarah Adams for gabbing, Julie Walters for being awesome, and, especially, Elena Cotto, whose Red Pen of Justice keeps me in line.

Steve Dawson was a great roommate, and became a prize of a friend. There were some long nights in the living room wondering out loud what the hell we could do with a PhD in astrophysics if we were to split for the real world. Thanks to Steve, now I know: have your weekends off, enjoy your life, and make money.

Thanks to Colleen Henry for being such a great friend. She has made my life extremely bright since I met her. She's going to be the best PhD social welfarologist in the world. And thanks to her family, and Darcy, who have treated me like one of their own. I really appreciate their kindness and love.

Thanks, I guess, to Mother, the stray cat from my hood, who brought seven kittens into my closet one hot afternoon last July. Five were adopted, and two, Choo Choo and Coco, still live at the hacienda with Mother, who decided being feral isn't where it's at. Not long after, another stray showed up at my door, when my old pal Matt/Buck Prentiss came to California to fill Steve's vacancy. His low-stress philosophy helped me maintain an even strain during some absurdly stressful times. And he's been a great help, reminding me to take a break now and then.

I had the great fortune to be able to spend many months of my life in Green Bank, West Virginia. While it *is* a little out of the way, the Green Bank staff made me feel right at home. I made a lot of friends down there, and miss them quite a bit. First of all, the most important people to thank are the cafeteria girls because they always brought a smile to my face and kept me from going hungry: Barbara Taylor, Amy Thompson, Sue Roberts, Ashley McCarty, Sylvia Warner, Shirley Riggsby, and Rosie Slaven. I always enjoy my long conversations with Max Gum and Daryl Shinaberry on the way from, then back to, Dulles. The telescope operators are top-notch. I really enjoyed all the conversations and the time that we spent together (cumulatively, over 1500 hours): Eric Knapp, Greg Monk, Kevin Gum, Dave Rose, Barry Sharp, and Donna Stricklin. Whenever I had a software problem that needed fixing, Bob Garwood, Paul Marganian, Amy Shelton, Mark Clark, and Melinda Mello were all quick to solve it. The scientific staff was incredibly helpful in assisting in the implementation of our extremely non-standard observing methods. In particular, Toney Minter, Dana Balser, Rick Fisher, Frank Ghigo, and Ron Maddalena all invested a lot of time helping me out. Roger Norrod and Rich Lacasse were invaluable in sorting out the heavy details. Thanks a ton to Sue Ann Heatherly and Bill Saxton for helping me get some of our results out into the real world. Shirley Curry, Sherry Sizemore, and Becky Warner are organizational experts and were always looking out for me. Our observations were aided by the organizational work of Bob Anderson, Pete Chestnut, and especially Carl Bignell, who does an extremely hard job extremely well. Our polarization work was only possible due to the full support of Phil Jewell and Richard Prestage. Some other Green Bank friends who made my stays more enjoyable include Jim Braatz, Mike Stennes, Mary Meeks, Maggie Morgan, Don Nelson, and Brian Mason, who came through for me in a big way by finding the IAU definition of Stokes V hidden in a time capsule behind a secret panel in the Charlottesville library.

I met Karen O'Neil on my first trip to Arecibo. Within the first hour, she became an invaluable telescope pal. Like an angel at the side, she moved to Green Bank just as I switched my observing from Puerto Rico to West Virginia. She, and her husband Paul, couldn't have been more helpful. In addition to solving a lot of telescope mysteries for me, Karen has also given me a lot of great professional advice. I really appreciate everything.

Sadly, the development of remote observing has ended my visits to Puerto Rico, but my observations have been, and still are, made easier by the help of Phil Perillat, Chris Salter, and Arun Venkataraman. Edith Alvarez, Willie Portalatin, and Norberto Despiau were all very helpful to me during my time in Arecibo.

There are a few people who certainly influenced my life path along the way, and I thank them here: Bill Barrett, Chet Raymo, Don Goldsmith, and Dave Cudaback. Most especially, I thank Richard Spillane, my physics teacher for two years at Boston Latin. He passed away far too early at the age of 51. His classes, which I never paid a dime for, set me on this course; it has been disheartening to me that not one of my college courses, which I'm still repaying loans for, was as inspirational as his.

Most importantly, I would like to thank my family. Without the love and support of my mother and father, Mary and Arthur Robishaw, I would not have been able to pursue a career doing something that I love. They gave me everything I could ever need, and I truly appreciate it. Thanks to Uncle Tim for having me over so often and keeping me well fed, and for hipping me to Coppa Mista, Deadwood, and the Sopranos. Thanks to Uncle Richie for letting me use his computer to observe and get work done when back in Boston. I wouldn't have made it this far without the love and support of my brother, Art; he's the best friend a guy could have. He's also the one who told me to take physics instead of chemistry. Above all, I thank my Nana, Lillian Cashman, who I miss very much. She isn't here to see this final product, but I know she's proud all the same.

Finally, as I finish the 24th grade, I thank the great decider, George W. Bush. There were many times over the last eight years when I thought that I might not have the right stuff to earn a PhD in astrophysics. During these dark times, George was always there for me, reminding me that if he could be President of the United States, then *anything* is possible. As we are both getting ready to leave office (mine is 751B Campbell Hall), I begin this dissertation by following his lead, and stealing his words:

> I'm going to try to see if I can remember as much to make it sound like I'm smart on the subject.

The argument in the past has frequently been a process of elimination: one observed certain phenomena, and one investigated what part of the phenomena could be explained; then the unexplained part was taken to show the effects of the magnetic field. It is clear in this case that, the larger one's ignorance, the stronger the magnetic field.

LODEWIJK WOLTJER, 1966

xvi

Chapter 1

Introduction

I hate quotations. Tell me what you know.

RALPH WALDO EMERSON, 1849

M agnetic fields play an important role in the structure and dynamics of the interstellar medium (ISM). The energy density of magnetic fields is roughly 1 eV cm^{-3}; this is identical to the energy density of cosmic rays, gas motions, starlight, and the cosmic microwave background. It is assumed that the first three are connected, and that the latter two are just a coincidence. However, it is clear that magnetic fields cannot be ignored in the study of the ISM.

Unfortunately, there is a difficulty in directly quantifying their importance observationally: they are notoriously hard to measure. Magnetic fields interact with matter in space in such a way that they leave a fingerprint on radiation that is either emitted or absorbed by this matter. This fingerprint is embedded in the polarization of the received astronomical radiation. For example, one can infer the magnetic field in the plane of the sky by measuring the polarization of starlight or dust emission. Synchrotron radiation also allows the observer to indirectly estimate the plane-of-sky field at the location where the radiation was generated.

A more direct observational method is that of Faraday rotation, which allows the observer to probe the line-of-sight component of the magnetic field. Partially linearly polarized radiation passing through an ionized medium that is immersed in a magnetic field will have its polarization position angle rotated by different amounts at different frequencies. One can measure the linear polarization at multiple frequencies and estimate the field strength. There are a number of caveats when using this method. First, Faraday rotation takes place inside the source where the radiation is generated. In addition, polarized sources are often variable, so one must be careful that multi-wavelength estimates of the polarization are measured relatively close in time. Most importantly, the inferred field strength is weighted by the electron density along the entire line of sight: this is not an unambiguously measurable quantity. There is, however, one magnetometer that directly probes the magnetic field at the source of radiation.

1.1 The Zeeman Effect: A Real Astrophysical Magnetometer

The Zeeman effect has become an invaluable tool for astronomers studying the magnetic field in our Galaxy. Various spectral lines will be split and polarized when originating in a region that is embedded in a magnetic field. This splitting will be directly proportional to the line-of-sight magnetic field strength. Unlike Faraday rotation, the Zeeman effect directly probes the magnetic field at the location where the spectral line is formed, and its spectral signature is not disturbed as the radiation passes through various interstellar media on its way to our telescope. It is, therefore, one of the extremely few in situ measurements in astrophysics, and certainly the most far-reaching.

The Zeeman effect was first used to measure astrophysical magnetic fields exactly 100 years ago. Exactly 40 years ago, radio astronomy joined the game. Since that time, a small group of dedicated observers has continued the search for magnetic fields in the ISM of the Milky Way via Zeeman splitting of radio frequency lines in emission and absorption. To date, detections have been made in the 21 cm transition of neutral hydrogen (H I), the 18 cm main and satellite lines of hydroxyl (OH), the 113 GHz hyperfine lines of cyanogen (CN), the 22 GHz rotational transition of water (H_2O) in masers, and most recently, 6.7 GHz methanol (CH_3OH) maser emission. A number of other transitions have been searched for unsuccessfully.

1.2 If It's So Amazing, Why Doesn't Anybody Use It?

While the Zeeman effect has been used successfully to probe the magnetic field in both diffuse and dense gas throughout the Milky Way (see Heiles & Crutcher 2005 for a thorough review), there have been far more non-detections than discoveries. The interstellar field strength is weak, on the order of a microgauss, so its Zeeman imprint in circularly polarized radiation takes many hours of integration to reveal itself. Moreover, the details of radio spectropolarimetry are extremely daunting for any newcomer. These considerations have not changed in the last 30 years, at which time Verschuur (1979) suggested:

> Apart from very long lasting and very tedious experimentation, it does not seem as if the area of 21 cm Zeeman searches will be a bonanza for eager graduate students seeking to break into radio astronomy in a dramatic way!

I was keenly aware of this when I began the research that has culminated in this dissertation. After observing my fair share of featureless Stokes V spectra, I began to suspect that I would be writing a thesis based solely on upper limits to magnetic field strengths in the Galaxy. However, with the accumulation of time and experience, and the deliverance of a large dose of good fortune, the Zeeman effect began to reveal secrets about magnetic fields, not only in our own Milky Way, but, in an unexpectedly copious and dramatic fashion, in galaxies far, far away.

1.3   Thesis Outline

In Chapter 2, we will discuss the fundamental physics of the Zeeman effect, and explain how it is measured using single-dish radio telescopes. We will describe a search for magnetic fields in the Taurus Molecular Cloud complex in Chapter 3. Chapter 4 will provide the details of the very first emission-line extragalactic Zeeman detections using OH megamaser emission in ultraluminous infrared galaxies. Chapter 5 describes the detection of an 80 μG field in a damped Lyα absorber at $z = 0.692$ toward 3C 286.

4

Chapter 2

Measuring the Zeeman Effect with a Single-Dish Radio Telescope

> How can one look happy when he is thinking about the anomalous Zeeman effect?

WOLFGANG PAULI, 1946

The Zeeman effect is one of the most powerful tools in the astronomer's shed. It is the only method that allows an observer to make an in situ measurement of the magnetic field strength at interstellar distances. It has been used copiously in optical astronomy, especially in studies of our own Sun, where field strengths of one to thousands of gauss can be probed. However, the techniques for measuring the Zeeman effect in single-dish radio astronomy involve a mastery of spectropolarimetry. Two orthogonal polarizations are received, and unlike in an optical system, where the two polarizations traverse the same path before detection, each of the orthogonal polarizations in a radio system traverses an independent path, experiencing unique gain and phase variations that are time-dependent. These two orthogonal polarizations must be calibrated and combined to produce a circular polarization spectrum, which is where the Zeeman effect leaves its spectral fingerprint. Unfortunately, the details of polarimetric calibration are not well-known in the radio community, and this extends to the staff of our major single-dish observatories, virtually guaranteeing that the study of single-dish radio Zeeman splitting will be confined to a small, familiar group. This is unfortunate, but as Tinbergen (2003) points out:

> Measurement of circular polarization is mostly looked upon as a specialist craft, applied by polarization freaks to answer uninteresting questions.

What could be less uninteresting than directly probing magnetic fields halfway across the Milky

Way, or possibly, halfway across the universe? Most radio astronomers disregard the polarization information that they are receiving and are only interested in averaging the signals from the two orthogonal feeds. With minimal effort, those radio astronomers interested in the polarization characteristics of the signal they are receiving could obtain these data. In order to see the spectral fingerprint of Zeeman splitting, it is essential to fully understand the details of spectropolarimetry and become a full-fledged polarization freak. In this chapter, we shall develop a detailed explanation of how one uses a single-dish radio telescope to measure Zeeman splitting, and hopefully demystify somewhat the process by which this is done. As an added incentive to treat the development with an extra degree of care, we found a number of places in the literature where discrepancies in sign, factors of two, and phases of $180°$ were evident; these discrepancies cause plenty of confusion, and suggest that a clear, detailed explanation is worthwhile.

In § 2.1, we present the history of the Zeeman effect. We provide a detailed explanation of how polarized radiation is represented in § 2.2. We then outline a physical derivation of the Zeeman effect and quanitfy its spectral signature in § 2.3. Finally, § 2.4 presents a lengthy and detailed explanation of how one observes the Zeeman effect with a single-dish radio telescope.

2.1 The Zeeman Effect

The Zeeman effect has a very rich history in physics and astronomy. In this section we outline its discovery and interpretation, and qualitatively explain how it is observed. Finally, we present a brief summary of its usefulness as an astronomical probe of magnetic fields.

2.1.1 Experiment and Theory Stop Confronting One Another, Win a Nobel Prize

Between 1890 and 1893, Pieter Zeeman (1865–1943) was a graduate student at the University of Leiden working in the laboratory of his advisor, Heike Kamerlingh Onnes.[1] His doctoral research involved the study of a magneto-optic effect, named after John Kerr, in which the plane of polarization of light is rotated when reflected from a magnetized surface. Zeeman interrupted his studies in an attempt to discover if the shape of spectral lines from a sodium flame would change when influenced by a magnetic field. No change was detected so he went back to finishing his thesis, *Measurements Concerning the Kerr Effect*, and earned his doctorate in 1893. He stayed on at Leiden and worked as an assistant to Kamerlingh Onnes and Hendrik A. Lorentz, a theoretician

---

[1]Who would later win the 1913 Nobel Prize for the production of liquid helium. Like Zeeman, he would eventually have an effect named after him: the Onnes effect is the ability of superfluid liquids to "creep" up surfaces.

investigating electromagnetism. In 1896 August, Zeeman decided to repeat his previously failed experiment, this time using improved experimental apparatus that included a more powerful 10 kG magnet and a higher-resolution grating. Zeeman (1897a) explains:

> Probably I should not have tried this experiment again so soon had not my attention been drawn some two years ago to the following quotation from Maxwell's sketch of Faraday's life. Here we read: "Before we describe this result we may mention that in 1862 he made the relation between magnetism and light the subject of his very last experimental work. He endeavoured, but in vain, to detect any change in the lines of the spectrum of a flame when the flame was acted on by a powerful magnet." If a Faraday thought of the possibility of the above-mentioned relation, perhaps it might be yet worth while to try the experiment again with the excellent auxiliaries of spectroscopy of the present time.[2]

On 1896 September 2, while observing the D_1 and D_2 doublet lines of sodium, he switched on his magnet and found that "the lines become wider until they are two to three times as wide." When he switched off the magnet, they instantly returned to their original width. He went to great experimental lengths to ensure that the change in the line profile (for both emission and absorption lines) was due solely to magnetic influence, and not to the temperature or density increase of the sodium.

Onnes presented Zeeman's publication of his results at the Dutch Academy of Sciences on 1896 October 31. Unable to understand what was causing this broadening, Zeeman approached Lorentz and asked him if the results could be interpreted in terms of Lorentz's 1892 electromagnetic model of charged particles. Only days after Zeeman had submitted his first publication (Kox 1997), Lorentz concocted a theoretical model for the broadening of the sodium lines based on his own theory. In this model, a charged particle, which he called an "ion," is harmonically bound to the center of an atom; the spectral lines of the sodium doublet would correspond to the oscillation frequencies of the ions. A force (later to be called the Lorentz force) would be applied to the ions when a magnetic field is turned on. His theory therefore predicted the following observables (see § 2.3.1 for the classical derivation):

1. The lines should be split in frequency by an amount directly proportional to the magnetic field: $\Delta\nu = eB/(4\pi m_e c)$, where e is the charge of the "ion," B is the field strength, m_e is the mass of the "ion," and c is the speed of light.

---

[2]It was "a Faraday" who discovered the very first magneto-optic phenomenon in 1845 when he found that the plane of polarization of light is rotated by a magnetic field whose direction is parallel to the direction of propagation. This would later be called the Faraday effect, and is now more commonly known as Faraday rotation.

2. When viewed along the magnetic field direction, each spectral line will be split into a doublet, with no emission at the original line frequency; the split lines will appear shifted from
   the original frequency by the frequency splitting given in 1 above and will be oppositely
   circularly polarized.

3. When viewed perpendicular to the field, each spectral line will be split into a triplet: one
   component will be seen at the original line frequency, and two split lines will appear at the
   shifted frequencies. All of the lines will be linearly polarized, with the unshifted component polarized along the magnetic field direction, and the two shifted components polarized
   perpendicularly to the field direction.

Since Zeeman did not observe an actual splitting of the lines, only a broadening, Lorentz pointed
out that the above polarization characteristics ought to be visible in the broadened wings of the
spectral lines (Zeeman 1897a):

> Prof. Lorentz, to whom I communicated these considerations, at once kindly informed
> me of the manner in which, according to his theory, the motions of an ion in a mag-
> netic field is to be calculated, and pointed out to me that, if the explanation following
> from his theory be true, the edges of the lines of the spectrum ought to be circularly
> polarized. The amount of widening might then be used to determine the ratio between
> charge and mass, to be attributed in this theory to a particle giving out the vibrations
> of light.
>
> The above-mentioned extremely remarkable conclusion of Prof. Lorentz relating
> to the state of polarization in the magnetically widened lines I have found to be fully
> confirmed by experiment.

Zeeman presented Lorentz's analysis and his own experimental results to the Dutch Academy of
Sciences on 1896 November 28. Zeeman's two initial publications were concatenated and revised,
and an English translation was published in both *The Philosophical Magazine* (Zeeman 1897a)
and the fifth volume of *The Astrophysical Journal* (ApJ; Zeeman 1897b).

Informed by Lorentz's theory, Zeeman now set out to observe a complete splitting of spectral
lines in a magnetic field. By increasing the field strength to 32 kG and using the "especially
sharp" blue 480 nm line of cadmium rather than sodium, the lines were completely separated. He
followed his discovery paper with a report entitled *Doublets and Triplets in the Spectrum Produced
by External Magnetic Forces* (Zeeman 1897c), in which he presents the discovery of the circularly
polarized doublet (sans the central component) when viewed longitudinally along the magnetic
field,[3] and a triplet of linearly polarized components when viewed transverse to the magnetic field.

---

[3]This was achieved by using a magnet with perforated poles.

Returning briefly to the original discovery paper (Zeeman 1897a), Lorentz had pointed out to Zeeman that the measurement of line broadening would allow for the determination of the charge-to-mass ratio. Zeeman's measured result was "of the order of magnitude of 10^7 electromagnetic c.g.s. units" (Zeeman 1897a). It is rarely mentioned that this measurement was made and published months before Sir Joseph J. Thomson's cathode ray determination, and is remarkably close to Thomson's more accurate measurement (Thomson 1897). Zeeman (1931), looking back on the measurement, reports:

> I found for the ratio of charge to mass the high value 10^7 c.g.s., quite different from the electrolytic ratio for hydrogen. I may be permitted to mention here that having come to this conclusion, I immediately went to Lorentz to tell him. He remarked: "That looks really bad; it does not agree at all with what is to be expected."[4]

As we shall see in § 2.1.2, while Lorentz's theory worked quite well for explaining the broadening of the sodium doublet, it wouldn't be long before experimental observations began to diverge from "what is to be expected." This experiment also provided the very first measurement of the sign of the ion's charge, which, of course, would later be named the electron. Zeeman (1897a) outlines:

> It may be deduced from the experiment of § 20 whether the positive or the negative ion revolves.
> If the lines of force were running toward the grating, the right-handedly circularly polarized rays appeared to have the greater period. Hence in connection with § 18, it follows that the positive ions revolve, or at least describe the greater orbit.

Here begins the first in a long, long line of errors (continuing to the present epoch) involving the sense of circular polarizations in Zeeman effect experiments (see $ 2.4.2). Zeeman (1897c) corrects himself a few months later, explaining:

> I must, however, correct my statement in § 24 of my former paper. I now see that if the lines of force are running toward the grating, the right-handedly circular polarized rays appear to have the greater period. Hence the radiation is due chiefly to the motion of a *negatively*-charged particle. Probably my mistake arose from a faulty indication of the axis of the $\frac{1}{4}\lambda$ plate used.

This was the first detection of the negative charge of Lorentz's ions. Later that year, Thomson (1897) came to associate cathode rays with "corpuscles" and measured their charge to be negative, but he failed to associate these corpuscles with the Zeeman-Lorentz ions; he went on to directly measure the charge and mass of the "corpuscle" in 1899. By the end of that year, the physics

---

[4]This should always be the expected response when an observer seeks advice from a theorist.

community had accepted that the Zeeman-Lorentz ions were Thomson's cathode-ray corpuscles and the particle was named the *electron*,[5] digging up a term introduced by George Stoney in 1891 (Pais 1986).

Zeeman's discoveries and Lorentz's interpretation and guidance were an exemplary marriage of experiment and theory, so much so that the two shared the Nobel Prize in 1902 (the second year of its existence) "in recognition of the extraordinary service they rendered by their researches into the influence of magnetism upon radiation phenomena."

2.1.2   A Tale of Two Zeeman Effects

> History has presented us with an unfortunate terminology: the *normal* pattern is rather anomalous, while the *anomalous* Zeeman pattern is quite normal.
>
> JERRY MARION, 1965

Not long after Zeeman's discovery of the triplet was explained by Lorentz, the theoretical framework came tumbling down. In 1897 December, Thomas Preston photographed quartets and sextets in the sodium D lines along the magnetic field direction. He also found that some triplets had spacings larger than could be explained with the classical treatment of Zeeman and Lorentz (Preston 1898). The former observation revealed a true irony of Zeeman's original discovery: had he been able to resolve the splitting of the sodium doublet, rather than just the broadening, Lorentz's theory would not have been supported by Zeeman's observations!

The physical reason for these deviations remained completely mysterious for the next two decades. As the magnetic field strengths began to increase, the situation only became more confounding, and Paschen & Back (1912) christened the new, classically unexplainable observations the *anomalous Zeeman effect*.[6] To differentiate this from the classically explained Zeeman triplet, the original case was termed the *normal Zeeman effect*. The mystery continued for many years throughout the development of the "old" quantum theory, which could not tackle the problem until a number of ad hoc suggestions were made by Landé (1921). Among these were the introduction of a factor g (which he originally called *speziellen Proportionalitätsfaktor* but which now, thankfully, bears his name); and the suggestion that the quantum number M must be able to take half-integer values—an idea so wild at the time that when a young, unpublished student named Werner Heisen-

---

[5]Interestingly, Lorentz began to use the term *electron* immediately while Thomson, who would later win the 1906 Nobel Prize "in recognition of the great merits of his theoretical and experimental investigations on the conduction of electricity by gases," refused to use the word *electron* for a number of years.

[6]Thanks to them we also have another term in the Zeeman taxonomy, the Paschen-Back effect, which is simply the Zeeman effect for large fields.

berg had independently come to the same conclusion, his advisor (Arnold Sommerfeld) forbade him from publishing it (Pais 1986), calling it "downright silly" (Forman 1970). However, Landé's formula did not *solve the problem* of what was causing the deviations from Zeeman's triplet: it simply expanded the mathematical framework of the theory to match the observations. Or as Pais (1986) opines:

> Landé's formula in fact demonstrates excellently how, in the days of the old quantum theory, gifted physicists were able to make important progress without quite knowing what they were doing.

However, progress was made when, in trying to interpret Landé's result, Wolfgang Pauli[7] determined that the anomalous Zeeman effect "is due to a peculiar not classically describable two-valuedness of the quantum theoretical properties of the valency electron" (Pauli 1925). While this paper generated the Nobel Prize-winning exclusion principle, Pauli himself did not take the next step; instead, it was Uhlenbeck & Goudsmit (1925) who identified the two-valuedness as the electron spin.

It took 28 years to develop an understanding of the anomalous Zeeman effect. At the end we were left with the paradox that the anomalous Zeeman effect is indeed quite normal since it applies to any atom with nonzero spin ($S \neq 0$) radiating in a magnetic field, while the normal Zeeman effect is only applicable to atoms with no net electronic spin ($S = 0$). Therefore, the normal Zeeman effect is truly the more anomalous Zeeman effect!

So where does that leave us? Well, considering that H I isn't even normal since $S \neq 0$, it would seem that we should have a lot of splitting lines to keep track of in radio astronomy. However, in the weak-field limit, the anomalous Zeeman effect appears as a triplet (more on this in § 2.3.2), and therefore looks like a normal Zeeman pattern! Barring a spectacular discovery, radio astronomers studying the interstellar magnetic field (with strengths ranging from micro- to milligauss) will always be in the weak-field limit,[8] and should fully expect to see the normal Zeeman pattern.

---

[7]Who, like Zeeman, also had an effect named after him: the *Pauli effect* was named for his ability to break experimental equipment simply by standing in the vicinity (Enz 2002).

[8]The anomalous Zeeman effect also degenerates into a normal triplet in the strong-field limit as well: the regime where the external field is comparable to the internal field for the atom or molecule is where the anomalous Zeeman effect provides its strangest splitting patterns.

2.1.3   The Zeeman Effect as an Astrophysical Magnetometer

> I decided to test the components of the spot doublets for
> evidence of circular polarization.

GEORGE ELLERY HALE, 1908

In 1908, George Ellery Hale was the first to measure astronomical Zeeman splitting by observing the polarization properties of spectral lines in sunspots, estimating the solar field to be "2900 gausses" (Hale 1908). The discovery of Zeeman splitting in other stars would have to wait almost 40 more years until 1946, when Horace Babcock[9] measured fields of $\sim$1500 G (Babcock 1947) in a half-dozen stars. It would take another 20 years before the Zeeman effect was brought to bear on white dwarf stars (Kemp 1970), in which field strengths were measured to exist in the megagauss–gigagauss range![10]

However, the interstellar magnetic field had completely eluded both optical and radio astronomers. Bolton & Wild (1957) pointed out that Zeeman splitting of the 21 cm hyperfine transition of H I occurring in a very weak field of $\sim$1 μG should be measurable in narrow absorption lines. This speculation set radio astronomers on a race to measure the interstellar field. The search continued for 10 years until 1968 July 4, when Verschuur (1968) unambiguously detected Zeeman splitting in the 21 cm line in absorption against the bright radio background source Cas A. Because 21 cm profiles are so broad and the typical field in the interstellar medium (ISM) is so weak, the Zeeman effect is only observable in H I as a *broadening*, just like Zeeman originally observed in 1896. However, in Galactic masers, where the line widths can be as small as 1 km s^{-1} and can probe high-density star-forming regions where the field can be 1000 times stronger than is typically observed in the diffuse ISM, the emission lines can be completely split by the Zeeman effect. It is worth noting that two years prior to the 21 cm detection of the Zeeman effect, the team of Davies et al. (1966), followed by Barrett & Rogers (1966), discovered Zeeman splitting in the circularly polarized "anomalous OH emission" (which was to later be understood as maser emission, a possibility that Barrett & Rogers suggest)[11] toward multiple sources including W3 and W49. However, both teams disavowed (the second team more strongly than the first) the certainty that Zeeman splitting was responsible for the circular polarization signatures, because the polarization spectra did not have quite the expected pattern for the Zeeman effect; we now know this complication is

---

[9]Interestingly, his father, Harold, also an astronomer, published many papers in the ApJ about laboratory Zeeman effect measurements from 1911–1929. Both were Bruce Medalists of the Astronomical Society of the Pacific.

[10]Above $B \approx 10^5$ G, a term in the Zeeman shift that is quadratic with B and depends on the fourth power of the principle quantum number begins to become important (Jenkins & Segrè 1939); luckily, we don't probe such high strengths in interstellar fields, so we won't be worrying about the quadratic term in radio measurements.

[11]At the time, OH maser emission was attributed to the magnificently named element "mysterium."

brought about because of complex radiative transfer effects in polarized maser emission.

Since the discovery of the Zeeman effect in the ISM, a small group of radio astronomers has focused its efforts on exploiting this astrophysical magnetometer by means of single-dish radio observations; this group has included, but is certainly not limited to, Gerrit Verschuur, Carl Heiles, Tom Troland, and Dick Crutcher.[12] For almost 40 years, Heiles and Troland have been studying the Galactic interstellar field in diffuse regions using Zeeman splitting of 21 cm emission and absorption lines,[13] and they have recently reported results from an extensive Arecibo survey that suggest a mean CNM magnetic field strength of 6 μG (Heiles & Troland 2005). Over a similar time scale, Crutcher and Troland have led an extensive single-dish campaign to measure Zeeman splitting in molecular lines, chiefly OH in absorption and emission. These searches have yielded measured magnetic field strengths of roughly a dozen microgauss in molecular clouds (Crutcher 1999; Troland & Crutcher 2008). Zeeman splitting of CN has also been detected in dense star-forming cores, where strong fields of half a milligauss are probed (Crutcher et al. 1999; Falgarone et al. 2008).

The Zeeman effect has also been detected in Galactic maser emission (see Vlemmings 2007 for a review) from OH ($\sim$1–40 mG) and H_2O ($\sim$15–650 mG). Recently, Vlemmings (2008) reported the detection of Zeeman splitting of 6.7 GHz CH_3OH methanol maser emission in 17 massive star-forming regions corresponding to magnetic fields of roughly 20 mG.

Kazes et al. (1991) were the first to see the Zeeman effect outside of the Galaxy. The field was measured in a 21 cm absorption feature of a high-velocity system around Per A, and the detection was confirmed by Sarma et al. (2005). We have used the Zeeman splitting of 1667 MHz emission from OH megamasers (see Chapter 4) in multiple starburst galaxies to detect fields of milligauss strength; these are the first Zeeman measurements to directly probe the magnetic field inside external galaxies. Finally, we climb the extragalactic distance ladder to report in Chapter 5 on the discovery of Zeeman splitting in redshifted 21 cm absorption from a dampled Lyα system at $z = 0.692$.

---

[12]The students of the latter two are now among the world's experts in Zeeman measurements using radio interferometry.

[13]Thomas H. Troland, like the author, was a graduate student of Carl Heiles. He gave up his chance at winning a Nobel Prize working with Joseph Taylor studying pulsars at Amherst (Huguenin et al. 1970) to come to Berkeley and work with Heiles from 1970–1980 on single-dish radio observations of Zeeman splitting in the 21 cm line. While the Troland (Td) is a unit of retinal illuminance, measured in cd m^{-2}, around the Berkeley Astronomy Department the Troland is the equivalent of a 10 year predoctoral stay, a unit the author has nearly achieved.

2.2   Representations of Polarized Radiation

> If light is man's most useful tool, polarized light is the
> quintessence of utility.
>
> WILLIAM SHURCLIFF, 1962

In order to understand how incoming radiation bearing the spectral signature of Zeeman splitting interacts with the telescope to produce the final data product that is analyzed by the astronomer,
it is essential that we review the details of the polarization of radiation. The technique of *spectropolarimetry* obviously involves a multiwavelength analysis of the polarization, but we start out
by considering only monochromatic radiation and generalize our results by adding frequency dependence later.

2.2.1   Monochromatic Plane-Polarized Radiation

The polarization of radiation is defined by the motion of the electric field vector as a function
of time within a plane perpendicular to the direction of propagation that contains the electric and
magnetic field vectors. This plane is known as the *plane of polarization* and the general shape that
the field traces out in this plane is an ellipse; not surprisingly, this is known as the *polarization
ellipse*. In order to represent polarization, we need to parametrize the behavior of the electric field.

Consider a monochromatic plane wave traveling in the $+\hat{\mathbf{z}}$ direction of a right-handed coordinate system. To follow astronomical convention, we shall orient the $+\hat{\mathbf{z}}$ direction to point from
the source toward the observer and rotate the xy plane to be aligned with the equatorial coordinate
system so that $+\hat{\mathbf{x}}$ is pointing toward the north and $+\hat{\mathbf{y}}$ is pointing toward the east.[14]

The components of the electric field along the x and y axes will be a function of time and
position along the z axis:

$$E_x(z, t) = E_{0x} \cos\left(kz - 2\pi\nu t + \delta_x\right) , \tag{2.1a}$$

$$E_y(z, t) = E_{0y} \cos\left(kz - 2\pi\nu t + \delta_y\right) , \tag{2.1b}$$

where k is the wave number, ν is the frequency of the monochromatic radiation measured in hertz,
t is the time measured in seconds, E_{0x} and E_{0y} are the amplitudes of each component measured in
volts per meter, and δ_x and δ_y are the phases of each component measured in radians. The electric
field vector is then given by:

$$\mathbf{E}(z, t) = E_x(z, t)\hat{\mathbf{x}} + E_y(z, t)\hat{\mathbf{y}} . \tag{2.2}$$

---

[14]This aligns the x axis with declination and the y axis with right ascension.

We can use Euler's formula, $e^{i\theta} = \cos\theta + i\sin\theta$ (where $i = \sqrt{-1}$), to represent the electric field and its components as complex numbers, the real parts of which correspond to the physical components of the electric field:

$$E_x(z,t) = E_{0x}e^{i(kz-2\pi\nu t+\delta_x)}\,, \tag{2.3a}$$

$$E_y(z,t) = E_{0y}e^{i(kz-2\pi\nu t+\delta_y)}\,. \tag{2.3b}$$

Since both components share the oscillatory term $\exp[i(kz-2\pi\nu t)]$, we can recast equation (2.2) as:

$$\mathbf{E}(z,t) = \mathbf{E_0}e^{i(kz-2\pi\nu t)} = (\mathcal{E}_x\hat{\mathbf{x}} + \mathcal{E}_y\hat{\mathbf{y}})\,e^{i(kz-2\pi\nu t)}\,, \tag{2.4}$$

where the quantity $\mathbf{E_0}$ is known as the *full Jones vector*[15] (Jones 1941) and is represented as a two-element column matrix whose components are the complex amplitudes:

$$\mathbf{E_0} = \begin{bmatrix} \mathcal{E}_x \\ \mathcal{E}_y \end{bmatrix} = \begin{bmatrix} E_{0x}e^{i\delta_x} \\ E_{0y}e^{i\delta_y} \end{bmatrix}\,. \tag{2.5}$$

At a given position z along the direction of propagation (let's take $z=0$ for simplicity), the tip of the electric field vector $\mathbf{E}$ will trace out an ellipse in time with orthogonal components given by:

$$E_x(t) = \mathcal{E}_x e^{-i(2\pi\nu t)}\,, \tag{2.6a}$$

$$E_y(t) = \mathcal{E}_y e^{-i(2\pi\nu t)}\,. \tag{2.6b}$$

These components define the previously mentioned polarization ellipse.

In polarization work, the *normalized Jones vector* is often used to ease calculations. It is obtained by dividing the full Jones vector by its total intensity, $\sqrt{E_{0x}^2 + E_{0y}^2}$, and ignoring the absolute phase (which should *not* be done if considering interference!). Therefore, we can define the relative phase to be:

$$\delta \equiv \delta_y - \delta_x\,, \tag{2.7}$$

and our resulting normalized Jones vector can be written as:

$$\mathbf{E_0} = \begin{bmatrix} a \\ be^{i\delta} \end{bmatrix}\,, \tag{2.8}$$

where a and b are just the normalized x and y amplitudes of the electric field, respectively. If $\delta = m\pi$, where m is an integer, then the radiation is linearly polarized; if m is a half integer and $a = b$, then the radiation is circularly polarized; otherwise it is elliptically polarized.

---

[15]There is likely no more thorough an exegesis on the matter of the Jones calculus than Kliger et al. (1990).

The major axis of the polarization ellipse will be oriented at an angle α with respect to the x axis where

$$\tan 2\alpha = \frac{2E_{0x}E_{0y}\cos\delta}{E_{0x}^2 - E_{0y}^2}; \qquad -\frac{\pi}{2} < \alpha \leq \frac{\pi}{2}, \tag{2.9}$$

with the ratio of the major to minor axis amplitudes, χ, being defined by

$$\sin 2\chi = \sin 2\alpha \sin\delta. \tag{2.10}$$

For linear polarization, $\delta = 0$ (or any integer multiple of π), and the electric field oscillates along a line oriented at angle α. Therefore, since α can only be measured modulo 180°, linear polarization has an *orientation*, not a *direction* (for which α could be measured modulo 360°).

We can now consider some specific examples of polarization using the normalized Jones vectors. For linearly polarized light, we find the general solution:

$$\mathbf{E_0}(\alpha) = \begin{bmatrix} \cos\alpha \\ \sin\alpha \end{bmatrix}, \tag{2.11}$$

where α is the orientation of the linear polarization. With respect to the xy coordinate system, we have horizontal and vertical linear polarizations given by:

$$\mathbf{E_0}(\alpha = 0) = \frac{1}{\sqrt{2}}\begin{bmatrix} 1 \\ 0 \end{bmatrix} \qquad \text{and} \qquad \mathbf{E_0}(\alpha = \tfrac{\pi}{2}) = \frac{1}{\sqrt{2}}\begin{bmatrix} 0 \\ 1 \end{bmatrix}. \tag{2.12}$$

These represent an orthonormal basis.[16]

Next we consider circularly polarized radiation. Let's consider $m = \frac{1}{2}$ for simplicity. If $\delta = +\frac{\pi}{2}$, then the horizontal component of the field leads the vertical component in time by 90°. If we consider what happens at $z = 0$, $E_x(t)$ would reach $+E_{x0}$ one quarter of a period before $E_y(t)$ would reach $+E_{y0}$, and the electric field would look to be rotating counterclockwise in the xy plane (the plane of polarization) as viewed by an observer who is located far away on the positive z axis. If $\delta = -\frac{\pi}{2}$, then $E_x(t)$ lags $E_y(t)$ by 90°, and the field would appear to be rotating clockwise to the observer. In radio astronomy, counterclockwise rotation ($\delta = +\frac{\pi}{2}$) of the electric vector as seen by the observer is called *right-handed circular polarization* (RHCP or RCP), while clockwise rotation ($\delta = -\frac{\pi}{2}$) of the electric vector as seen by the observer is called *left-handed circular polarization* (LHCP or LCP).

Here lies a real danger in circular polarimetry. There is no good reason to attach the right- and left-handed labels to either clockwise or counterclockwise rotation; it is simply a matter of definition. *All physics texts present the handedness of circular polarization in the opposite manner from*

---

[16]Two vectors, $\mathbf{A}$ and $\mathbf{B}$, form an orthonormal basis if their inner product is zero ($\mathbf{A} \cdot \mathbf{B}^* = 0$) and the inner product of each with itself is unity ($\mathbf{A} \cdot \mathbf{A}^* = \mathbf{B} \cdot \mathbf{B}^* = 1$).

that which we have defined, and this is simply because physicists have had a long tradition (Born & Wolf 1999) of attaching RCP to counterclockwise rotation as seen by an observer with the light approaching. It must be said that optical astronomers use the classical physics definition of RCP and LCP and therefore are not in agreement with radio astronomers, who have unfailingly used the Institute of Electrical and Electronics Engineers (IEEE) definition (IEEE 1997) of RCP and LCP since at least 1942 (more on this in § 2.4.2). In many astronomical radiation and polarization texts, this leads to the presentation of the classical physics definitions of RCP and LCP, since most are written by optical astronomers (e.g., Rybicki & Lightman 1979; Leroy 2000; del Toro Iniesta 2003; Landi degl'Innocenti & Landolfi 2004). It should also be pointed out that the description of whether $E_x(t)$ leads or lags $E_y(t)$ will depend on two definitional statements: whether the phase of the plane wave has been defined as $(kz - 2\pi\nu t)$ or $(2\pi\nu t - kz)$, and whether the relative phases have been defined as $\delta = \delta_y - \delta_x$ or $\delta = \delta_x - \delta_y$. Clearly, when it comes to circular polarization, one must take care to define one's terms.

Having considered the details of circular polarization, we can show from equation (2.5) that the Jones vectors can be written:

$$\mathbf{E_0}(\text{RCP}) = \frac{1}{\sqrt{2}} \begin{bmatrix} 1 \\ i \end{bmatrix} \quad \text{and} \quad \mathbf{E_0}(\text{LCP}) = \frac{1}{\sqrt{2}} \begin{bmatrix} 1 \\ -i \end{bmatrix}. \tag{2.13}$$

The inner product of these vectors is zero, so they are orthogonal and can be used to represent all other forms of polarization. This is an important point because the feeds of radio telescopes are designed to probe either orthogonal linear or circular polarizations. Therefore, since either orthogonal linears or orthogonal circulars represent a basis, we can represent any kind of polarization with either. For instance, linear polarization at $\alpha = 45°$ can be represented in the linear basis or the circular basis:

$$\mathbf{E_0}(\alpha = \tfrac{\pi}{4}) = \frac{1}{\sqrt{2}} \begin{bmatrix} 1 \\ 1 \end{bmatrix} \tag{2.14}$$

$$= \frac{1}{\sqrt{2}} \left(\begin{bmatrix} 1 \\ 0 \end{bmatrix} + \begin{bmatrix} 0 \\ 1 \end{bmatrix} \right) = \frac{1}{\sqrt{2}} \mathbf{E_0}(\alpha = 0) + \frac{1}{\sqrt{2}} \mathbf{E_0}(\alpha = \tfrac{\pi}{2}) \tag{2.15}$$

$$= \frac{1+i}{2\sqrt{2}} \begin{bmatrix} 1 \\ i \end{bmatrix} + \frac{1-i}{2\sqrt{2}} \begin{bmatrix} 1 \\ -i \end{bmatrix} = \frac{e^{-i\pi/4}}{\sqrt{2}} \mathbf{E_0}(\text{RCP}) + \frac{e^{+i\pi/4}}{\sqrt{2}} \mathbf{E_0}(\text{LCP}). \tag{2.16}$$

In practice, dual orthogonal linear feeds will have a small amount of circular response, and dual orthogonal circular feeds will have a small amount of linear response. Therefore, we briefly discuss the most general polarization state, elliptical polarization, which can be represented as a linear combination of linear and circular polarizations. Since this is the general form of polarization, it is worth stating the form of the normalized Jones vector for right-handed elliptical

TABLE 2.1

NORMALIZED JONES AND STOKES VECTORS FOR SIMPLE POLARIZATION STATES

Polarization State (1)	α (2)	δ (3)	$\mathbf{E_o}$ (4)	$\mathbf{E_o}$(CP) (5)	S (6)
Linear Horizontal	$0°$	$\cdots$	$\begin{bmatrix} 1 \\ 0 \end{bmatrix}$	$\frac{1}{\sqrt{2}}\begin{bmatrix} 1 \\ 1 \end{bmatrix}$	$\begin{bmatrix} 1 \\ 1 \\ 0 \\ 0 \end{bmatrix}$
Linear Vertical	$90°$	$\cdots$	$\begin{bmatrix} 0 \\ 1 \end{bmatrix}$	$\frac{1}{\sqrt{2}}\begin{bmatrix} i \\ -i \end{bmatrix}$	$\begin{bmatrix} 1 \\ -1 \\ 0 \\ 0 \end{bmatrix}$
Linear at $\alpha = +45°$	$+45°$	$0°$	$\frac{1}{\sqrt{2}}\begin{bmatrix} 1 \\ 1 \end{bmatrix}$	$\frac{1}{2}\begin{bmatrix} 1+i \\ 1-i \end{bmatrix}$	$\begin{bmatrix} 1 \\ 0 \\ 1 \\ 0 \end{bmatrix}$
Linear at $\alpha = -45°$	$-45°$	$0°$	$\frac{1}{\sqrt{2}}\begin{bmatrix} 1 \\ -1 \end{bmatrix}$	$\frac{1}{2}\begin{bmatrix} 1-i \\ 1+i \end{bmatrix}$	$\begin{bmatrix} 1 \\ 0 \\ -1 \\ 0 \end{bmatrix}$
Linear at any α	α	$0°$	$\begin{bmatrix} \cos\alpha \\ \sin\alpha \end{bmatrix}$	$\frac{1}{\sqrt{2}}\begin{bmatrix} e^{-i\alpha} \\ e^{+i\alpha} \end{bmatrix}$	$\begin{bmatrix} 1 \\ \cos 2\alpha \\ \sin 2\alpha \\ 0 \end{bmatrix}$
Right-Handed Circular (RCP) . . .	$\cdots$	$+90°$	$\frac{1}{\sqrt{2}}\begin{bmatrix} 1 \\ i \end{bmatrix}$	$\begin{bmatrix} 1 \\ 0 \end{bmatrix}$	$\begin{bmatrix} 1 \\ 0 \\ 0 \\ 1 \end{bmatrix}$
Left-Handed Circular (LCP)	$\cdots$	$-90°$	$\frac{1}{\sqrt{2}}\begin{bmatrix} 1 \\ -i \end{bmatrix}$	$\begin{bmatrix} 0 \\ 1 \end{bmatrix}$	$\begin{bmatrix} 1 \\ 0 \\ 0 \\ -1 \end{bmatrix}$

polarization (RHEP or REP) and left-handed elliptical polarization (LHEP or LEP):

$$\mathbf{E_0}(\text{REP}) = \frac{1}{\sqrt{a^2 + b^2 + c^2}} \begin{bmatrix} a \\ b - ic \end{bmatrix} \quad \text{and} \quad \mathbf{E_0}(\text{LEP}) = \frac{1}{\sqrt{a^2 + b^2 + c^2}} \begin{bmatrix} a \\ b + ic \end{bmatrix}. \quad (2.17)$$

where $E_{x0} \equiv a$, $E_{y0} \equiv \sqrt{b^2 + c^2}$, and $\tan \delta = c/b$.

We list the normalized Jones vectors for various states of polarization in Table 2.1. Column (4) lists the normalized Jones vector in the linear basis, $\mathbf{E_0}$, and column (5) lists the normalized Jones vector in the circular polarization basis, $\mathbf{E_0}(\text{CP})$

Unfortunately, Jones vectors and the polarization ellipse cannot be used to represent unpolarized or partially polarized radiation. They are extremely useful tools for understanding the telescope response to polarized radiation, but since there is no such thing as a completely polarized astronomical source, we can't even use Jones vectors *some of the time* to represent the emission from a polarized astronomical source. Luckily, there exists a perfect tool for measuring partially polarized radiation: the Stokes parameters.

2.2.2 The Stokes Parameters and Stokes Vector

In 1852,[17] Sir George G. Stokes developed a set of four parameters that completely quantify the propagation of polarized radiation (Stokes 1852). Now known as *Stokes parameters*, they are formally defined (Rybicki & Lightman 1979; Goldstein 2003; Landi degl'Innocenti & Landolfi 2004) as:

$$I \equiv \langle E_x E_x^* \rangle + \langle E_y E_y^* \rangle , \quad (2.18a)$$

$$Q \equiv \langle E_x E_x^* \rangle - \langle E_y E_y^* \rangle , \quad (2.18b)$$

$$U \equiv \langle E_x E_y^* \rangle + \langle E_x^* E_y \rangle , \quad (2.18c)$$

$$V \equiv i \left(\langle E_x E_y^* \rangle - \langle E_x^* E_y \rangle \right) , \quad (2.18d)$$

where the angle brackets denote a time average of the electric field, and the superscripted asterisks denote the complex conjugate. The Stokes parameters can be calculated in any orthogonal basis; here we have chosen the Cartesian coordinate system in the plane of polarization, xy. Here, we are representing the Stokes parameters as functions of the electric field. In radio astronomical measurements, the feeds convert the incident electric fields to voltages; the detection process produces power proportional to the square of these voltages. The raw data product is usually given in an arbitrary correlator analog-to-digital unit, which is then converted via calibration to a temperature

---

[17]Four years before Maxwell first published a paper on electromagnetism (Shu 1991).

unit (kelvins) or a flux density unit (millijanskys). The Stokes parameter spectra are then presented in these calibrated units. Therefore, there is an understood constant on the right-hand side of each Stokes parameter accounting for intensity calibration.

The order of terms is critical in equation (2.18d); in particular, this definition requires the definition of the relative phase to be that of equation (2.7). If the order were reversed so that $\delta = \delta_x - \delta_y$, then we would need to replace each product in brackets in equations (2.18) with its complex conjugate. It should be evident that this does not affect Stokes I, Q, or U, but will cause Stokes V to be opposite in sign from the definition of equation (2.18d); one can then use the property $-i = i^{-1}$ to rewrite equation (2.18d) as $iV \equiv \left\langle E_x E_y^* \right\rangle - \left\langle E_x^* E_y \right\rangle$ (after Heiles 2002 and Rybicki & Lightman 1979). In addition, this correction would need to be made if equation (2.7) *were* used, but the phase of the plane wave was defined with the form $(\omega t - kz)$ rather than $(kz - \omega t)$ (Landi degl'Innocenti & Landolfi 2004).

The angle brackets in equations (2.18) are used to represent the time average of the complex field products. This is required because we are treating the polarized radiation as quasi-monochromatic. There is no such thing as purely monochromatic radiation,[18] and only monochromatic light can be fully polarized. Therefore, the electric field vector for quasi-monochromatic light, which can only be partially polarized, will not trace out an exact ellipse with time. But the elliptical nature of the partial polarization will be recovered if the products are averaged over a time that is long relative to the wave period. Since the shortest integration time available for most correlators is 1 s, the averaging time is long enough for the polarization properties to be probed by the radio receiver.

By substituting equations (2.6) into equations (2.18), we can produce the more commonly found representation[19] of the Stokes parameters:

$$
\begin{aligned}
I &= E_{0x}^2 + E_{0y}^2\,, \\
Q &= E_{0x}^2 - E_{0y}^2\,, \\
U &= 2E_{0x}E_{0y}\cos\delta\,, \\
V &= 2E_{0x}E_{0y}\sin\delta\,.
\end{aligned}
\tag{2.19}
$$

From equations (2.19) and equation (2.9), it can be seen that the angle that the polarization ellipse

---

[18] Just considering the Fourier transform of a pure monochramatic delta function, it would require an infinite amount of time and an infinite amount of space to create such radiation.

[19] Optics, radiation, and astronomy texts usually provide this set of Stokes parameters, and will often include their representation as a function of the polarization ellipse parameters.

makes with the horizontal can be expressed by:

$$\tan 2\alpha = \frac{U}{Q}. \tag{2.20}$$

The parameter I simply represents the total intensity of the radiation; it does not store any polarization information. The parameter Q can be thought of as the tendency for linear polarization to be aligned horizontally rather than vertically. If $Q > 0$ there is an excess of polarized radiation along the horizontal, while for $Q < 0$, there is an excess of vertically polarized light. Likewise, the U parameter is the tendency for the linear polarization to be aligned at $+45°$ to the horizontal, with $U < 0$ meaning an excess in linear polarization at an angle $-45°$ to the horizontal. Finally, the Stokes V parameter is a direct measure of the circular polarization, and is the difference between the right-handed circular polarization and the left-handed circular polarization. For positive Stokes V, there is an excess of RCP over LCP when using the IEEE conventions. Since V is proportional to the sine of the relative phase δ, RCP ($0 < \delta < \pi$) corresponds to positive V, and it is reiterated that, should we reverse our definitions of the relative phase to $\delta \equiv \delta_x - \delta_y$ or of the absolute phase to $(\omega t - kz)$, the sense of Stokes V will change. We will revisit the details of circular polarization conventions in § 2.4.2 The Stokes vector is listed for simple polarization states in Table 2.1.

The four Stokes parameters are commonly arranged in a 1×4 column matrix known as the *Stokes vector*[20]:

$$S = \begin{bmatrix} I \\ Q \\ U \\ V \end{bmatrix}. \tag{2.21}$$

We point out that the Stokes vector, like the Jones vector, is often normalized to ease computations: this is accomplished by dividing the vector by Stokes I. We will not differentiate between the normalized and unnormalized cases.

The Stokes vector is crucially important in polarimetry because it allows us to represent partially polarized radiation, something that the Jones calculus will not allow. One can decompose the Stokes vector into a totally unpolarized part and an elliptically polarized part:

$$S = \begin{bmatrix} I - \sqrt{Q^2 + U^2 + V^2} \\ 0 \\ 0 \\ 0 \end{bmatrix} + \begin{bmatrix} \sqrt{Q^2 + U^2 + V^2} \\ Q \\ U \\ V \end{bmatrix}. \tag{2.22}$$

---

[20]To be pedantic, we should be careful to not confuse two of the possible meanings of the word *vector*. While the Jones vector is a 1×2 column matrix and is therefore a vector in the sense defined by linear algebra, it is also a spatial vector in the sense defined by physics and vector calculus: the components of the Jones vector are the magnitudes of the electric field in an orthonormal basis. However, the Stokes vector is a vector only in the linear algebra sense: the four component parameters do not form a basis.

The degree of polarization, or fractional polarization, is the ratio of the total power of the polarized emission to the total power:

$$P = \frac{I_{\text{pol}}}{I_{\text{tot}}} = \frac{\sqrt{Q^2 + U^2 + V^2}}{I}; \qquad 0 \leq P \leq 1. \tag{2.23}$$

It should be stressed that the Jones vector complements the Stokes vector in the sense that the former can be employed to represent the combination of coherent beams of radiation, while the latter cannot.

Having presented the details of how to represent quasi-monochromatic partially polarized radiation, we shall use these concepts in § 2.3.1 to describe the expected spectropolarimetric fingerprint of the Zeeman effect, and again throughout § 2.4 to illuminate the details of the measurement of polarized radio spectra.

2.2.2.1   Nomenclature

Most physics and electrical engineering texts define the Stokes parameters using the set of variables $\{S_0, S_1, S_2, S_3\}$. But astronomers are never satisfied with keeping things simple; instead, we use the variables $\{I, Q, U, V\}$ to represent the Stokes parameters. Why? The answer seems to have gone to the grave with Subrahmanyan Chandrasekhar. Almost 100 years after their initial development, the Stokes parameters were resurrected for astronomical use when Chandrasekhar (1947) was developing the equation of radiative transfer for scattering in stellar atmospheres. He recalls (Chandrasekhar 1989) discovering them one afternoon when he "took down from the library shelves the collected papers of Rayleigh, Kelvin, and Stokes," and instantly found the parameters $\{A, B, C, D\}$ defined by Stokes (1852). Chandrasekhar comments:

> That was written in 1852; but no book on optics that had been written for one hundred years after Stokes (with one exception, as I came to know later) had included an account of the topic, and no physicist whom I had consulted seemed even to be aware of the problem.[21]

However, the Stokes parameters had been rediscovered and actively used in optical physics for almost twenty years before Chandrasekhar's rediscovery. It seems that Soleillet (1929) was the first to dredge up Stokes's invention and he proposed (en Français) that the set be relabeled $\{I, M, C, S\}$. These terms were very descriptive: I for the total *intensity* of the radiation; M

---

[21]The one exception that Chandrasekhar refers to is that of Walker (1904) who uses the notation $\{P, Q, R, S\}$, which at the very least is alphabetical if not all that descriptive. Given his disavowal of knowledge of Walker's work, we draw attention to Chandrasekhar's apparently *random* selection of Q as the second parameter.

for *minus* (or, more appropriately, *moins*, where $M \equiv A_x^2 - A_y^2$, in his notation), for the difference between two linear polarizations; C for *cosine* ($C \equiv 2A_x A_y \cos\phi$); and S for *sine* ($S \equiv 2A_x A_y \sin\phi$).[22] This treatment was elaborated in English by Perrin (1942). Chandrasekhar (1947) also made the obvious choice of I to represent the total *intensity*, but there is no obvious explanation for why Chandrasekhar chose to represent the linear polarization parameters by Q and U, or why he chose V to represent the circular polarization parameter. Despite its seemingly arbitrary selection, the set $\{I, Q, U, V\}$ has become standard in astronomical usage, and was introduced to radio astronomy by Cohen (1958).

2.3   A Physical Description of the Normal Zeeman Effect

The quantum mechanical derivation of the Zeeman effect is carried through in its hairy glory in a number of texts; the most thorough are Condon & Shortley (1957), Landi degl'Innocenti & Landolfi (2004), and White (1934). There are very few places where the classical derivation of the normal Zeeman effect is reproduced, likely because, as we pointed out, the normal Zeeman effect is really not all that commonly encountered. Rather, in a laboratory environment, one is more likely to need the full quantum details of the atomic or molecular anomalous Zeeman pattern that one is studying. For our purposes, it will suffice to present a simple classical derivation in § 2.3.1, and to follow this in § 2.3.2 with a few details concerning the Zeeman effect that can only be explained by quantum mechanics.

2.3.1   Classical Treatment

> It is of great advantage to the student of any subject to read the original memoirs on that subject, for science is always most completely assimilated when it is in the nascent state.
>
> JAMES CLERK MAXWELL, 1873

There are a number of ways to approach the problem, from the consideration of a single harmonically bound electron, to the consideration of a full distribution of atoms radiating in a magnetic field. These various approaches will all lead to differing coefficients in the total power radiated by the source. What is important is not the absolute power radiated, but an understanding

---

[22]The set $\{I, M, C, S\}$ (and later $\{I, Q, U, V\}$) corresponds to the original Stokes parameters in the order $\{A, C, D, B\}$. Also, note that if one shifts the polarization basis to orthogonal circulars, then Stokes S in the $\{I, M, C, S\}$ scheme becomes RCP *minus* LCP (and does not contain a *sine* term), thereby potentially causing some definitional confusion since the M term no longer contains the *minus* when represented in this basis. Perhaps Chandrasekhar was prescient enough to give us polarized parameter descriptors that were equally undescriptive in each basis!

of the polarization properties of the split lines and their relative intensities. A very detailed and complex derivation can be found in Landi degl'Innocenti & Landolfi (2004), while the most simple derivations can be found in Marion (1965), Sommerfeld (1954), and, of course, Zeeman (1897a). Here we will closely follow the development by Goldstein (2003), which considers the radiation from a single electron that is harmonically bound to its nucleus.

If the electron is moving at nonrelativistic velocities, the electric field of the accelerating charge is given by Jackson (1998) equation (14.18):

$$\mathbf{E}(\mathbf{x}, t) = \frac{e}{4\pi\varepsilon_0 c^2} \frac{\hat{\mathbf{n}} \times (\hat{\mathbf{n}} \times \dot{\mathbf{v}})}{R}, \tag{2.24}$$

where e is the charge of the electron, ε_0 is the permittivity of free space, c is the speed of light, R is the distance from the atom to the observer, $\hat{\mathbf{n}}$ is the unit vector pointing from the electron to the observer, and $\dot{\mathbf{v}}$ is the vector acceleration of the electron. The triple cross product can be rewritten $\hat{\mathbf{n}} \times (\hat{\mathbf{n}} \times \dot{\mathbf{v}}) = \hat{\mathbf{n}}(\hat{\mathbf{n}} \cdot \dot{\mathbf{v}}) - \dot{\mathbf{v}}$. Expressing the acceleration in a Cartesian coordinate system, which we take to be centered out in space at the position of the atom to which this particular electron is bound, we can write:

$$\dot{\mathbf{v}} = \ddot{x}\hat{\mathbf{x}} + \ddot{y}\hat{\mathbf{y}} + \ddot{z}\hat{\mathbf{z}}, \tag{2.25}$$

where we'll take the $+\hat{\mathbf{z}}$ direction to be aligned with the direction of an external magnetic field $\mathbf{B}$ at this position in space. We are so far from the atom that we can take $\hat{\mathbf{n}} = \hat{\mathbf{r}}$, such that the triple cross product becomes

$$\hat{\mathbf{n}} \times (\hat{\mathbf{n}} \times \dot{\mathbf{v}}) = \hat{\mathbf{r}}(\hat{\mathbf{r}} \cdot \dot{\mathbf{v}}) - \dot{\mathbf{v}}. \tag{2.26}$$

We can write the unit vectors in the Cartesian coordinate system as functions of the unit vectors in the spherical coordinate system

$$\hat{\mathbf{x}} \equiv \sin\theta\cos\phi\,\hat{\mathbf{r}} + \cos\theta\cos\phi\,\hat{\boldsymbol{\theta}} - \sin\phi\,\hat{\boldsymbol{\phi}}, \tag{2.27a}$$

$$\hat{\mathbf{y}} \equiv \sin\theta\sin\phi\,\hat{\mathbf{r}} + \cos\theta\sin\phi\,\hat{\boldsymbol{\theta}} + \cos\phi\,\hat{\boldsymbol{\phi}}, \tag{2.27b}$$

$$\hat{\mathbf{z}} \equiv \cos\theta\,\hat{\mathbf{r}} - \sin\theta\,\hat{\boldsymbol{\theta}}, \tag{2.27c}$$

and use these to represent the acceleration of equation (2.25) in terms of the spherical unit vectors. Substituting into equation (2.26) we can write:

$$\hat{\mathbf{r}}(\hat{\mathbf{r}} \cdot \dot{\mathbf{v}}) - \dot{\mathbf{v}} = -\hat{\boldsymbol{\theta}}(\ddot{x}\cos\theta\cos\phi + \ddot{y}\cos\theta\sin\phi - \ddot{z}\sin\theta) + \hat{\boldsymbol{\phi}}(-\ddot{x}\sin\phi + \ddot{y}\cos\phi), \tag{2.28}$$

but we can see that this term is symmetric in ϕ. This suggests that one's azimuthal position in the xy plane doesn't matter: we can observe the radiating charge at any angle around the magnetic field direction. Thus, we can arbitrarily set $\phi = 0$ to be toward the observer. Substitution into equation (2.24) yields the radiation field equations for an accelerating electron:

$$E_\theta = \frac{e}{4\pi\varepsilon_0 c^2 R} \left[\ddot{x}(t)\cos\theta - \ddot{z}(t)\sin\theta\right] , \tag{2.29a}$$

$$E_\phi = \frac{e}{4\pi\varepsilon_0 c^2 R} \left[\ddot{y}(t)\right] . \tag{2.29b}$$

Equations (2.29) describe the general field for an accelerating electron.

Let's now discuss the physical model of an electron that is bound to an atom that is in the presence of an external magnetic field (which we've already modeled to be aligned with our z axis). This is the exact model that Lorentz proposed in 1896. The model allows us to write the equation of motion for the electron as:

$$m_e\ddot{\mathbf{r}} + k\mathbf{r} = -e\left(\mathbf{v}\times\mathbf{B}\right) , \tag{2.30}$$

where m_e is the mass of the electron, k is an analog of the spring constant that accounts for the restoring force, and of course, the right-hand side of the equation is just the Lorentz force acting on the electron due to the magnetic field $\mathbf{B}$. It is then straightforward to show that the Cartesian components of this force can be reduced to:

$$\ddot{x} + \omega_0^2 x = -\left(\frac{e\mathbf{B}}{m_e}\right)\dot{y} , \tag{2.31a}$$

$$\ddot{y} + \omega_0^2 y = -\left(\frac{e\mathbf{B}}{m_e}\right)\dot{x} , \tag{2.31b}$$

$$\ddot{z} + \omega_0^2 z = 0 , \tag{2.31c}$$

where we have defined $\omega_0 = (k/m_e)^{1/2}$ to be the natural frequency of the oscillating electron. We find the solutions for this system of second order differential equations to be:

$$x(t) = A\cos\omega_L t \,\cos\omega_0 t , \tag{2.32a}$$

$$y(t) = A\sin\omega_L t \,\cos\omega_0 t , \tag{2.32b}$$

$$z(t) = A\cos\omega_0 t , \tag{2.32c}$$

where A is the amplitude of the electron's oscillation about the atom, and we have defined the Larmor precession frequency[23] to be:

$$\omega_L \equiv \frac{e}{2mc}\mathbf{B} . \tag{2.33}$$

---

[23] As suggested by its name, this is the frequency with which an electron will precess about the direction of a magnetic field.

We express the trigonometric terms in complex exponential form, defining the two frequencies $\omega_\pm \equiv \omega_0 \pm \omega_{\rm L}$; after differentiating twice with respect to time, we find:

$$\ddot{x}(t) = -\frac{A}{2} \left[\omega_+^2 \exp(i\omega_+ t) + \omega_-^2 \exp(i\omega_- t) \right] , \qquad (2.34a)$$

$$\ddot{y}(t) = i\frac{A}{2} \left[\omega_+^2 \exp(i\omega_+ t) - \omega_-^2 \exp(i\omega_- t) \right] , \qquad (2.34b)$$

$$\ddot{z}(t) = -A\omega_0^2 \exp(i\omega_0 t) . \qquad (2.34c)$$

Substituting equations 2.34 into equations (2.29) we find that the electrical field generated from the accelerating harmonically bound electron is:

$$E_\theta = \frac{eA}{8\pi\varepsilon_0 c^2 R} \left[\cos\theta \left\{ \omega_+^2 \exp(i\omega_+ t) + \omega_-^2 \exp(i\omega_- t) \right\} + 2\omega_0^2 \exp(i\omega_0 t) \right] , \qquad (2.35a)$$

$$E_\phi = \frac{ieA}{8\pi\varepsilon_0 c^2 R} \left[\omega_+^2 \exp(i\omega_+ t) - \omega_-^2 \exp(i\omega_- t) \right] . \qquad (2.35b)$$

To parallel the derivation of Goldstein (2003), we originally considered the magnetic field direction to be pointing in the $+\hat{z}$ direction; this established the angle θ as a measure of the observer's position relative to the magnetic field direction. Recall that astronomical convention requires that a positive magnetic field point away from the observer. Thus, $\theta = 0$ corresponds to a negative field, while $\theta = 180°$ corresponds to a positive field.

The Stokes parameters of equations (2.18) are written for the orthogonal Cartesian basis, but they can be rewritten for the spherical coordinate system by replacing (x, y) with (ϕ, θ):

$$I \equiv \left\langle E_\phi E_\phi^* \right\rangle + \left\langle E_\theta E_\theta^* \right\rangle , \qquad (2.36a)$$

$$Q \equiv \left\langle E_\phi E_\phi^* \right\rangle - \left\langle E_\theta E_\theta^* \right\rangle , \qquad (2.36b)$$

$$U \equiv \left\langle E_\phi E_\theta^* \right\rangle + \left\langle E_\phi^* E_\theta \right\rangle , \qquad (2.36c)$$

$$V \equiv i \left(\left\langle E_\phi E_\theta^* \right\rangle - \left\langle E_\phi^* E_\theta \right\rangle \right) . \qquad (2.36d)$$

We can now substitute equations (2.35) into equations (2.36) and arrange the results into a general Stokes vector for the Zeeman effect:

$$S = \frac{8}{3} \left(\frac{eA}{8\pi\varepsilon_0 c^2 R} \right)^2 \begin{bmatrix} \frac{1}{4} \left(\omega_+^4 + \omega_-^4 \right) (1 + \cos^2\theta) + \frac{1}{2}\omega_0^4 \sin^2\theta \\ -\frac{1}{4} \left(\omega_+^4 + \omega_-^4 \right) \sin^2\theta + \frac{1}{2}\omega_0^4 \sin^2\theta \\ 0 \\ \frac{1}{2} \left(\omega_+^4 - \omega_-^4 \right) \cos\theta \end{bmatrix} . \qquad (2.37)$$

Since our derivation was for a single radiating electron, the terms to the left of the matrix are not especially meaningful; we'll absorb these factors into a single constant. Moreover, this single electron radiating in the presence of a magnetic field produces power that is proportional to the fourth power of the frequency, and the frequency distribution is a single delta function at each of the three frequencies ν_0, ν_-, and ν_+. The source of the radiation will, of course, be an ensemble of electrons bound to atoms, and the spectral lines will therefore have a frequency profile that is determined by a number of mechanisms, including natural, Doppler, and collisional broadening (Rybicki & Lightman 1979). The width of the broadened profile is typically much greater than the Zeeman splitting for typical ISM field strengths. In radio astronomy, the temperature profile of a spectral line centered at ν_0 is usually represented as $T(\nu - \nu_0)$, where the ν^4 dependence and the profile shape are incorporated into the function $T(\nu - \nu_0)$. We will consider the spectral signature of the Zeeman effect in terms of the Stokes I profile, which we label $I(\nu - \nu_0)$. Therefore, we will have profiles centered on each of the three Zeeman frequencies: $I(\nu - \nu_0)$, $I(\nu - \nu_-)$, and $I(\nu - \nu_+)$. We can now recast equation (2.37), absorbing the constant term into the profile shapes:

$$
S = \begin{bmatrix}
\frac{1}{4}\left\{I(\nu - \nu_+) + I(\nu - \nu_-)\right\}(1 + \cos^2\theta) + \frac{1}{2}I(\nu - \nu_0)\sin^2\theta \\
-\frac{1}{4}\left\{I(\nu - \nu_+) + I(\nu - \nu_-)\right\}\sin^2\theta + \frac{1}{2}I(\nu - \nu_0)\sin^2\theta \\
0 \\
\frac{1}{2}\left\{I(\nu - \nu_+) - I(\nu - \nu_-)\right\}\cos\theta
\end{bmatrix}.
\tag{2.38}
$$

As a sanity check, we can consider the case of radiation from a region where there is no magnetic field at the emitting source. In this case, since $B = 0$, there is no Zeeman splitting and $\nu_+ = \nu_- = \nu_0$. Equation (2.38) reduces to:

$$
S = I(\nu - \nu_0)\begin{bmatrix} 1 \\ 0 \\ 0 \\ 0 \end{bmatrix},
\tag{2.39}
$$

which is the Stokes vector for a single completely unpolarized spectral line centered at ν_0. It is clear that $I(\nu - \nu_0)$ is the Stokes I spectrum. We can now decompose equation (2.38) into three separate column vectors by grouping terms associated with the three independent frequencies:

$$
S = \frac{1}{4}\left(I(\nu - \nu_-)\begin{bmatrix} 1 + \cos^2\theta \\ -\sin^2\theta \\ 0 \\ -2\cos\theta \end{bmatrix} + I(\nu - \nu_0)\begin{bmatrix} 2\sin^2\theta \\ 2\sin^2\theta \\ 0 \\ 0 \end{bmatrix} + I(\nu - \nu_+)\begin{bmatrix} 1 + \cos^2\theta \\ -\sin^2\theta \\ 0 \\ 2\cos\theta \end{bmatrix} \right).
\tag{2.40}
$$

We immediately see that the spectral signature of the Zeeman effect is elliptically polarized emission at the split frequencies and pure linear polarization at the unsplit frequency.

Next, we consider what is observed when a magnetic field is oriented longitudinally to the line of sight and pointing toward the observer ($B < 0$), so that $\theta = 0°$ (this has come to be known

TABLE 2.2

MAGNETIC FIELD DIRECTION DETERMINED FROM THE ZEEMAN EFFECT

Field Direction[a]	$\nu_0 - \Delta\nu_z$	$\nu_0 + \Delta\nu_z$	Stokes V Profile[b]
Emission Lines			
$B > 0$	RCP	LCP	
$B < 0$	LCP	RCP	
Absorption Lines			
$B > 0$	LCP	RCP	
$B < 0$	RCP	LCP	

[a] A positive magnetic field points away from the observer by convention.
[b] V = RCP − LCP by IAU convention (IAU 1974).

as the *longitudinal Zeeman effect*):

$$S = \frac{1}{2} \left(I(\nu - \nu_-) \begin{bmatrix} 1 \\ 0 \\ 0 \\ -1 \end{bmatrix} + I(\nu - \nu_+) \begin{bmatrix} 1 \\ 0 \\ 0 \\ 1 \end{bmatrix} \right) . \tag{2.41}$$

It is clear that only circularly polarized emission will be seen. The component shifted to higher frequency will have positive intensity and will display RCP; the component shifted to lower frequency will have negative intensity and will display LCP. If the shift is much smaller than the line width of the spectral line, then the line will not be split: rather, as Zeeman first discovered, the line will be broadened, and the circular polarization signature implied from the Stokes vector will be that of the letter S on its side. This classic Zeeman shape is known as the "S curve" or the "circular Zeeman pattern." If the field is pointing away from the observer ($\theta = 180°$, $B > 0$) then, as seen in equation (2.40), the intensity of each component swaps sign and the S curve is flipped.

It is important to note that the sense of circular polarizations will be reversed for an absorption profile. This is sometimes called the *inverse Zeeman effect*. Consider unpolarized continuum radiation from a background source that is incident upon a gas cloud. This gas is in the presence of a magnetic field, which has a nonzero line-of-sight component that is pointing toward us. The Zeeman effect will cause the gas to selectively absorb the RCP component of the background emission at the higher frequency shift, thereby transmitting the LCP emission only; at the lower frequency shift, the LCP emission is absorbed and only the RCP is transmitted.[24] The net effect of this is to reverse the sign of the Stokes V S curve.

Table 2.2 shows the expected S curve for both emission and absorption as a function of increasing frequency. It should be noted that one must flip each curve when considering a spectrum plotted as a function of increasing velocity. Table 2.2 assumes the IEEE definition of RCP and LCP, and the IAU definition of Stokes $V \equiv \text{RCP} - \text{LCP}$; there will be much more about these conventions is § 2.4.2. Therefore, if one is using this table to infer the magnetic field direction from a Stokes V spectrum that is plotted as $\text{LCP} - \text{RCP}$, then one must also flip each curve.

Next, we consider observing the Zeeman effect when the magnetic field is in the plane of the sky so that $\theta = 90°$. This has come to be known as the *transverse Zeeman effect*. We see that equation (2.40) reduces to:

$$S = \frac{1}{4}\left(I(\nu - \nu_-)\begin{bmatrix} 1 \\ -1 \\ 0 \\ 0 \end{bmatrix} + 2I(\nu - \nu_0)\begin{bmatrix} 1 \\ 1 \\ 0 \\ 0 \end{bmatrix} + I(\nu - \nu_+)\begin{bmatrix} 1 \\ -1 \\ 0 \\ 0 \end{bmatrix}\right). \qquad (2.42)$$

It is clear that no circular polarization will be seen, only linear polarization. The intensity of the unshifted line is twice that of the split lines and is aligned horizontally ($Q > 0$), while the split lines are both aligned vertically ($Q < 0$). This is simply a result of our choice to align our viewing angle at $\phi = 0$ so that the plane-of-sky projection of the magnetic field would be aligned along the x axis (or declination axis). The general forms for Stokes Q and U will be a function of the parallactic angle (measured from north through east) at which the plane-of-sky component of the magnetic field is aligned.

Finally, we highlight a common notation used for the spectral features seen in the longitudinal and transverse Zeeman effects. The circular components seen at the shifted frequencies are often referred to as "σ components," with the higher-frequency component sometimes labeled σ^+ and the lower-frequency sometimes labeled σ^-. For the transverse Zeeman effect, the linear polarization feature at the unsplit frequency is called the "π component." Most physics texts fail to explain the origin of this extremely mysterious terminology. Some sources from the first half of the twentieth century do not use σ or π, but instead label the radiation that is observed perpendicular to the field as "s-radiation," and that seen longitudinally along the field as "p-radiation." The origin of these terms comes from the German words *parallel* (to wit, parallel) and *senkrecht* (perpendicular). The switch to the less directly parsable Greek representations π and σ had begun

---

[24]Note also that the linear polarization for each Zeeman component in an absorption profile will be orthogonal to its emission orientation.

by the time the old quantum theory was being developed. Landé (1921) had begun to use σ and π in his description of the anomalous Zeeman effect for which, as we have seen, a single line can be split into multiple σ and π components.

2.3.2 Quantum Considerations

> When Otto Stern measured the proton moment in the early 1930s, he was advised not to bother—elementary theory proved the result would be one nuclear magneton. Fortunately, Stern had a healthy disregard for elementary theory.

> DANIEL KLEPPNER

A detailed quantum mechanical derivation of the Zeeman effect shows that the frequency shift of either of the split lines (the σ components) from the unsplit frequency is proportional to the Larmor precession frequency multiplied by a proportionality factor:

$$\Delta\nu_Z = g\nu_L = \frac{geB}{4\pi m_e c} = \frac{g\mu_B}{h} B \,, \tag{2.43}$$

where the proportionality factor g is known as the Landé g-factor (Landé 1921) or the gyromagnetic factor for pure LS coupling. Its value is given by:

$$g = 1 + \frac{J(J+1) - L(L+1) + S(S+1)}{2J(J+1)} \,, \tag{2.44}$$

where S is the total spin of the atom or molecule, L is the total orbital angular momentum, and $J = L + S$ is the total angular momentum. It can easily be seen that spinless atoms or molecules, for which $S = 0$ and therefore $J = L$, have $g = 1$ and will display the classical normal Zeeman effect. The quantity $e/(4\pi m_e c)$ in equation (2.43) is known as the "Zeeman displacement" and is equivalent to the Bohr magneton, μ_B, divided by Planck's constant, h; it has a measured value of 1.39962418(42) Hz μG^{-1} (Cox 2000). The Zeeman displacement gives the shift of either of the σ components from the unsplit frequency; in radio Zeeman work, the Zeeman splitting is often given as the total frequency separation of the two σ components. This viewpoint introduces a factor known as the "splitting coefficient," which is defined to be the product of the g-factor and twice the Zeeman displacement: $b \equiv 2g\mu_B/h = 2.799249209(70)g$ Hz μG^{-1}. Substituting into equation (2.43), we find that the σ components will be shifted by:

$$\Delta\nu_Z = \frac{b}{2} B \,. \tag{2.45}$$

In general, even for atoms and molecules with $S \neq 0$, if the external field is weak compared to the internal field, then a normal triplet will be observed with the appropriate g-factor. In the strong-field limit, a triplet is also seen but L and S completely decouple and the details of the splitting are more complex. This strong-field Zeeman effect is also known as the Paschen-Back effect. In the intermediate regime, where the internal magnetic field is comparable to the external field, the anomalous Zeeman effect can cause the observed pattern of σ and π components to become very complex. Condon & Shortley (1957, ch. 16) provide an incredibly detailed derivation of the anomalous Zeeman effect in atoms, where one needs to consider separately the g-factors for both the initial and final transition levels. Townes & Schawlow (1975, ch. 11) address the even more complex case of calculating g-factors and Zeeman splitting for molecules. Molecular g-factors are often only experimentally determinable.[25]

Atoms and molecules that have unpaired electrons in the outer shells are known as "free radicals." A single O atom has two unpaired electrons in its outer shell, while a single H atom has only one; therefore the diatomic molecule OH has only one unpaired electron since the outer shell of the O shares the lone H electron. Free radicals with only a single unpaired electron are most important for the Zeeman effect and include H I, OH, and CN, all of which have been observed in the ISM.

Perhaps the most interesting tracer for radio astronomers is the Zeeman splitting of the 21 cm line, which is a hyperfine transition. Because the initial and final transition levels are brought about by an interaction with the atomic nucleus, one must multiply the electronic Landé g-factor by the term (Townes & Schawlow 1975):

$$\frac{F(F+1) + J(J+1) - I(I+1)}{2F(F+1)} , \tag{2.46}$$

where I is the nuclear spin quantum number, J is the total angular momentum without the nuclear spin, and F is the total angular momentum quantum number. Nafe & Nelson (1948) provide the detailed calculation of Zeeman splitting for the 21 cm line. Since the line originates from the $^2S_{1/2}$ state of hydrogen, which has $J = \frac{1}{2}$, $S = \frac{1}{2}$, and $L = 0$, the standard electronic Landé g-factor is $g = 2$. This must be multiplied by the nuclear factor in equation (2.46). For this transition, $J = \frac{1}{2}$ and $I = \frac{1}{2}$, and we find that the nuclear factor is also equal to $\frac{1}{2}$, yielding a total g-factor of $g = 1$. Therefore, for the 21 cm hyperfine transition, $b = 2.80$ Hz μG^{-1} and the splitting pattern will be a normal Zeeman triplet.

---

[25]For molecules with filled shells, such that there is no electronic angular momentum ($S = 0$; e.g., the H_2O molecule), the Zeeman splitting becomes proportional to the nuclear magnetic moment μ_n rather than the electron magnetic moment. The two are related by the ratio of the mass of the nucleus to that of the electrons, and hence $\mu_n = \mu_B/1836$.

2.3.3 Possible Zeeman Tracers

Heiles et al. (1993) compile the splitting coefficients for a number of atomic and molecular transitions; this reference is very difficult to obtain and is no longer in print, so we have partially reproduced their Table 1, and we present the results in our Table 2.3. The g-factor (and hence b) can be calculated theoretically, as for the 21 cm line, but for complex atomic and molecular transitions it must be measured experimentally in the laboratory. Some of the g-factors for these transitions have been updated by laboratory measurements since the publication of Heiles et al. (1993), so we include the latest values in Table 2.3. Finally, we mention that the g-factors are calculated as an intensity-weighted averge over the σ components for each sense of polarization if the Zeeman pattern is an anomalous multiplet, i.e., not a normal Zeeman triplet. For instance, the 1612 and 1720 MHz satellite lines of OH are split into six σ components (Davies 1974), and the value of $b = 1.31$ Hz μG^{-1} is an intensity-weighted average over the three σ components on either side of the unshifted frequency. Observationally, however, only two of the σ components are seen in Galactic maser emission, one in RCP and the other in LCP. Fish et al. (2003) point out that these are likely to be only the strongest two of the circular sextet, which suggests that the splitting coefficients for the 1612 and 1720 MHz transitions should probably be 0.800 Hz μG^{-1}, the coefficient for only the strongest σ component.

2.4 The Observational Measurement of the Zeeman Effect

> One can learn astronomy from books without ever looking through a telescope or handling a photographic plate, but that is not the Berkeley way.
>
> RON BRACEWELL, 1955

In § 2.3 we characterized the expected polarization pattern of spectral line radiation from an astronomical source in the presence of a magnetic field. In this section, we discuss the details of actually measuring such a signal. We begin at the end, so to speak, by discussing how one uses the Stokes V circular polarization spectrum to infer the magnetic field in a source. We then follow this with a discussion of how this partially polarized astronomical radiation is detected by our single-dish radio telescope, and then outline the calibration process for polarization observations.

TABLE 2.3

SPLITTING COEFFICIENTS FOR ZEEMAN TRANSITIONS

Species	Transition	ν (GHz)	b (Hz μG^{-1})	Ref.
Atomic Transitions				
H I	$^2S_{1/2}, F = 1\text{--}0$	1.420406	2.80	1
H I	H$n\alpha$ recombination lines	$\cdots$	2.80	1
C II	C$n\alpha$ & C$n\beta$ recombination lines	$\cdots$	2.80	2
Molecular Transitions				
CH	$^2\Pi_{3/2}, J = 3/2, F = 2\text{--}2$	0.701677	1.96	1
CH	$^2\Pi_{3/2}, J = 3/2, F = 1\text{--}1$	0.724788	3.27	1
OH	$^2\Pi_{3/2}, J = 3/2, F = 1\text{--}2$	1.6122	1.308	1
OH	$^2\Pi_{3/2}, J = 3/2, F = 1\text{--}1$	1.6654	3.270	1
OH	$^2\Pi_{3/2}, J = 3/2, F = 2\text{--}2$	1.6673	1.964	1
OH	$^2\Pi_{3/2}, J = 5/2, F = 2\text{--}1$	1.7205	1.308	1
OH	$^2\Pi_{3/2}, J = 5/2, F = 2\text{--}3$	6.0167	0.678	1
OH	$^2\Pi_{3/2}, J = 5/2, F = 2\text{--}2$	6.0307	1.582	1
OH	$^2\Pi_{3/2}, J = 5/2, F = 3\text{--}3$	6.0350	1.132	1
OH	$^2\Pi_{3/2}, J = 5/2, F = 3\text{--}2$	6.0490	0.678	1
CH$_3$OH	$J_N = 5_1\text{--}6_0 A^+$	6.668512	0.0011	8
C$_4$H	$N = 1\text{--}0, J = 3/2\text{--}1/2, F = 0\text{--}1$	9.493061	-2.457	3
C$_4$H	$N = 1\text{--}0, J = 3/2\text{--}1/2, F = 1\text{--}2$	9.497616	0.897	3
CCS	$J_N = 1_0\text{--}0_1$	11.119446	0.813	5
SO	$J_N = 1_2\text{--}1_1$	13.044000	1.93	6
OH	$^2\Pi_{3/2}, J = 7/2, F = 3^+\text{--}3^-$	13.434637	1.06	4
OH	$^2\Pi_{3/2}, J = 7/2, F = 4^+\text{--}4^-$	13.441417	0.795	4
C$_4$H	$N = 2\text{--}1, J = 5/2\text{--}3/2, F = 1\text{--}2$	19.014720	1.30	1
C$_4$H	$N = 2\text{--}1, J = 5/2\text{--}3/2, F = 2\text{--}3$	19.015144	0.93	1
CCS	$J_N = 2_1\text{--}1_0$	22.344033	0.767	5
SO	$J_N = 1_0\text{--}0_1$	30.001630	1.740	5
CCS	$J_N = 3_2\text{--}2_1$	33.751374	0.702	5
CCS	$J_N = 4_3\text{--}3_2$	45.379033	0.629	5
SO	$J_N = 2_1\text{--}1_0$	62.931731	1.379	5
SO	$J_N = 2_2\text{--}1_1$	86.094	0.47	1
C$_2$H	$N = 1\text{--}0, J = 3/2\text{--}1/2, F = 2\text{--}1$	87.316925	1.40	1
SO	$J_N = 3_2\text{--}2_1$	99.299875	1.043	5
CN	$N = 1\text{--}0, J = 1/2, F = 1/2\text{--}3/2$	113.14434	2.18	7
CN	$N = 1\text{--}0, J = 1/2, F = 3/2\text{--}1/2$	113.17087	-0.31	7
CN	$N = 1\text{--}0, J = 1/2, F = 3/2\text{--}3/2$	113.19133	0.62	7
CN	$N = 1\text{--}0, J = 3/2\text{--}1/2, F = 3/2\text{--}1/2$	113.48839	2.18	7
CN	$N = 1\text{--}0, J = 3/2\text{--}1/2, F = 5/2\text{--}3/2$	113.49115	0.56	7
CN	$N = 1\text{--}0, J = 3/2\text{--}1/2, F = 1/2\text{--}1/2$	113.49972	0.62	7
CN	$N = 1\text{--}0, J = 3/2\text{--}1/2, F = 3/2\text{--}3/2$	113.59972	1.62	7
SO	$J_N = 4_3\text{--}3_2$	138.178548	0.800	5
SO	$J_N = 5_4\text{--}4_3$	178.605168	0.634	5
CN	$N = 2\text{--}1, J = 3/2, F = 3/2\text{--}5/2$	226.3325	2.6	1

References.—(1) Heiles et al. (1993) and references therein; (2) Greve & Pauls (1980); (3) Turner & Heiles (2006); (4) Güesten et al. (1994); (5) Shinnaga & Yamamoto (2000); (6) Uchida et al. (2001); (7) Crutcher et al. (1996); (8) Vlemmings 2008.

2.4.1 Splitting Fitting

We saw in § 2.3 that the spectral signature of Zeeman splitting should be a characteristic S curve in the Stokes V spectrum. In practice, one estimates the line-of-sight magnetic field strength by least-squares fitting the functional form of that S curve, which we shall see is simply the derivative of the Stokes I spectrum, to the Stokes V spectrum. We also saw in equation (2.38), that the expected Zeeman pattern should include linear polarization at the unsplit line frequency as well as each of the split frequencies.

First, we consider the circular polarization. Since the symmetric definition of a derivative is:

$$\frac{df(x)}{dx} = \lim_{h \to 0} \frac{f(x+h) - f(x-h)}{2h},$$

we see that

$$\frac{dI(\nu - \nu_0)}{d\nu} = \lim_{\Delta\nu_z \to 0} \frac{I(\nu - [\nu_0 + \Delta\nu_z]) - I(\nu - [\nu_0 - \Delta\nu_z])}{2\Delta\nu_z},$$

so that for small splitting we can rewrite the Stokes V parameter from equation (2.40) as:

$$V(\nu) = \frac{dI}{d\nu} \Delta\nu_z \cos\theta. \tag{2.47}$$

It is standard to include a term proportional to the Stokes I spectrum on the right-hand side of equation (2.47). This term accounts for any gain difference between the orthogonal polarizations: a scaled version of the I spectrum will be present in V if the gains are slightly different.

2.4.1.1 Following the Factor of Two

The derived line-of-sight magnetic fields are extracted from the Stokes V spectrum according to equation (2.47). However, since the Zeeman displacement $\Delta\nu_z$ is directly proportional to the magnetic field, the B field is not explicitly expressed in this equation. Therefore, the equation can be rephrased to be explicitly dependent on the magnetic field strength. One of the first places that this equation was explicitly written[26] is Troland & Heiles (1982a), who state:

$$V(\nu) = \left(2.8 \text{ Hz } \mu\text{G}^{-1}\right) B \cos\theta \frac{dT(\nu)}{d\nu}, \tag{2.48}$$

where Heiles (1996) later clarifies that Troland & Heiles (1982a) took $T(\nu) = I(\nu)/2$; i.e., they fit the derivative of the average of the orthogonal polarizations, $T(\nu)$, to the Stokes V spectrum in

---

[26]Prior to this, all radio Zeeman effect papers contain a statement along the lines of: "the circular polarization spectrum has the form of the differential of the observed line profile."

order to derive the line-of-sight component of the magnetic field.[27] Since $b = 2.80$ Hz μG^{-1} for the 21 cm line of H I, we therefore have a general form of equation (2.48):

$$V(\nu) = \frac{d(I(\nu)/2)}{d\nu} bB \cos\theta \qquad (2.49)$$

$$= bB_\parallel \frac{dT(\nu)}{d\nu} . \qquad (2.50)$$

Heiles et al. (1993) provide the relation in equation (2.50) in their thorough review article on magnetic field observations. We can substitute equation (2.45) into equation (2.49) to find the form we derived in equation (2.47):

$$V(\nu) = \frac{dI(\nu)}{d\nu} \Delta\nu_z \cos\theta . \qquad (2.51)$$

It is clear that equations (2.48)–(2.51) are all equivalent forms, and that equations (2.49) & (2.51) provide two ways of looking at the same problem:

1. If one is using the Zeeman shift, $\Delta\nu_z$, which is the shift of either σ component from the unshifted central frequency, then one must calculate the derivative of the Stokes I spectrum.

2. If one is using the splitting coefficient, b, which gives the total frequency shift between the two σ components (and therefore twice that of the Zeeman shift), then one must calculate the derivative of the average of the orthogonal polarizations, $\frac{dT(\nu)}{d\nu}$, or divide the derivative of the Stokes I spectrum by two. Of course, this can also be viewed as taking the derivative of Stokes I, and dividing the splitting coefficient by two.

The factor of two present in equation (2.49) caused momentary lapses of certainty among some of the world's leading Zeeman experts when queried by the author, so it seems that making this point explicitly here is not just an exercise in completeness.

2.4.1.2 Extracting the Line-of-Sight Magnetic Field

There are two methods that are commonly used to extract the line-of-sight magnetic field from the circular polarization spectrum. Both involve least-squares fitting the derivative of the Stokes I spectrum to the Stokes V spectrum. The first method assumes that a single field is responsible for creating the Zeeman splitting across the entire line profile. This has come to be called the

---

[27] In radio astronomy, one sometimes encounters a redefinition of Stokes I to that of the *average* of two orthogonal polarizations, rather than their sum. This is truly an unacceptable practice and should be strongly discouraged. This probably originated with the tendency of radio astronomers to plot the line spectrum $T(\nu)$ as the average of the orthogonal polarizations. In this case $I(\nu) = 2T(\nu)$. It is more appropriate, and less potentially harmful, to state that one is using $T(\nu) \equiv I(\nu)/2$, than to *redefine* Stokes I.

"single-field fit" (Heiles 1988). This is usually all that is required if a single S curve is seen in the Stokes V profile, no matter how complex the Stokes I profile. It is perhaps a minor point, but the classical theory of least squares (Chauvenet 1960) assumes that the independent variable (the derivative of Stokes I in this case) has no uncertainty associated with it; obviously this assumption does not hold for the single-field fit.

The second method is typically used for a Stokes V profile with complex structure, for which one is usually not able to perform a single-field fit. The solution is to decompose the Stokes I profile into Gaussian components; then, one uses the noiseless derivative of the composite fitted profile in equation (2.51). This is more in the spirit of the classical method of least squares, because the derivative of the modeled Stokes I profile is a noiseless quantity. This will allow one to fit for a different field in each component; since each component is at a different velocity, and possibly a different line-of-sight distance, one can find very different field strengths in a single Stokes V spectrum. The caveat for this method is that a Gaussian decomposition of a spectral line profile is completely subjective; often, there is no physical motivation for the component choices.

Finally, before one proposes to observe the Zeeman effect, it is useful to know how long one will need to integrate before a detectable Stokes V signal will be apparent. Troland (1990) offers an empirically determined detection formula:

$$\sigma_{B_{||}} \approx 0.8\nu_0^{1/2} \left(\frac{d}{b}\right) \left(\frac{T_{\text{sys}}}{T_L}\right) \left(\frac{\Delta v}{\text{km s}^{-1}}\right)^{1/2} \left(\frac{\tau}{\text{hr}}\right)^{-1/2} \mu\text{G}, \qquad (2.52)$$

where ν_0 is the unsplit line frequency in GHz, b is the splitting coefficient in Hz μG^{-1}, T_{sys} is the system temperature, T_L is the line temperature, Δv is the FWHM velocity extent of the line in km s^{-1}, τ is the total integration time measured in hours, and the factor d represents the degradation in noise by the digitization of the signal by a correlation spectrometer (this factor is ≥ 1). This formulation is for a receiver that observes two orthogonal polarizations simultaneously (this is usually the case; we know of at least one receiver, the IRAM 30 m circa 1996, which had received only a single linear polarization [Crutcher et al. 1996], but which now appears to have a dual-linear feed [Falgarone et al. 2008]). In Zeeman work, a derived line-of-sight magnetic field in excess of $3\,\sigma_{B_{||}}$ is considered a detection, and often only a $2\,\sigma_{B_{||}}$ result is cause for joy.

2.4.1.3   Estimating the Plane-of-Sky Field via the Zeeman Effect

In general, unless the magnetic field at the emission or absorption source is completely aligned with the line of sight, the incoming polarization of the σ components will be elliptical, and that of the π component will be purely linear. In principal, one might then be able to measure the linear polarization of the Zeeman pattern and estimate the field strength and orientation in the plane of the sky. For a native dual-linear feed, the linear polarization is probed via Stokes Q and U. The symmetric definition of a second derivative is:

$$\frac{d^2 f(x)}{dx^2} = \lim_{h \to 0} \frac{f(x+h) - 2f(x) + f(x-h)}{h^2},$$

which, for small Zeeman splitting, allows us to rewrite the Stokes Q parameter in equation (2.40) as:

$$Q(\nu) = -\frac{1}{4}\frac{d^2 I}{d\nu^2}\Delta\nu_z^2 \sin^2\theta. \tag{2.53}$$

The general form of the linear Stokes parameters can be inferred from Table 2.1 to be dependent on twice the position angle, ϕ, of the plane-of-sky magnetic field:

$$Q(\nu) = -\frac{1}{4}\frac{d^2 I}{d\nu^2}\Delta\nu_z^2 \sin^2\theta \cos 2\phi, \tag{2.54}$$

$$U(\nu) = -\frac{1}{4}\frac{d^2 I}{d\nu^2}\Delta\nu_z^2 \sin^2\theta \sin 2\phi. \tag{2.55}$$

We note that this plane-of-sky position angle dependence differs from that of Crutcher et al. (1993), who set $Q \propto (\cos\phi - \sin\phi)$ and $U \propto \sin\phi$.[28] As we discovered in § 2.2.2, Q quantifies the excess horizontal linear polarization, and is therefore positive when the polarization is aligned more with the horizontal axis (north–south) than with the vertical (east–west). Considering the π component, which has its linear polarization aligned along the magnetic field orientation, we therefore find $Q > 0$ for $\phi = 0°$ and $\phi = 180°$, while $Q < 0$ for for $\phi = 90°$ and $\phi = -90°$. The relations for Crutcher et al. (1993) incorrectly quantify $Q < 0$ at $\phi = 180°$ and $Q > 0$ at $\phi = -90°$. A similar problem exists for their Stokes U.

While a careful consideration of the linear polarization in the small-splitting Zeeman effect allows us to put our development of polarimetric representations into practice, the second derivative of the Stokes I profile is extremely small. Therefore, it is not possible to obtain sufficient integration time to achieve the signal-to-noise ratio necessary to estimate the plane-of-sky field strength. In principle, the plane-of-sky field strength *is* measurable for maser emission, since the line widths are so narrow that the Zeeman effect will completely split the σ components from the π component. However, in practice, the interpretation of linear polarization in OH masers is very complex due to radiative transfer effects that are not completely understood. This can lead to linear polarization only being detected in one or two of the triplet components; according to Fish & Reid (2006), only one linearly polarized Zeeman triplet has ever been detected.[29]

---

[28] In their eq. (1), a factor of $\sin^2\theta$ is missing from the second equality in their definitions of both Q and U.
[29] Also, the unsplit π component is rarely ever seen to be completely linearly polarized in OH masers.

2.4.2   The Proper Definition of Stokes V and the Calibration of Circular Polariza-tion

> When signs are involved, even two Dutch physicists as
> scrupulous as Lorentz and Zeeman may make errors.
> _____
>   EMILIO SEGRÈ, 1976

Since the Zeeman effect is detected using the Stokes V spectrum, a precise definition of this parameter is necessary in order to correctly assign the *direction* of the probed line-of-sight magnetic field. Unfortunately, as we have seen in § 2.2.1, when it comes to circular polarization there is plenty of room for confusion if conventions are not explicitly defined, and astronomers have taken full advantage of this for at least the last 40 years to thoroughly obfuscate the situation. Of course, as mentioned in § 2.1.1, Zeeman himself started things off in 1896 when he measured a *positive* charge for the electron by confusing the sense of his observed circular polarizations.

To begin with, we reiterate the point that radio astronomers use the IEEE convention for circular polarization sense. The official IEEE definition (IEEE 1997) of a right-handed polarized wave is:

> **right-hand polarized wave:** A circularly or an elliptically polarized electromagnetic wave for which the electric field vector, when viewed with the wave approaching the observer, rotates counter-clockwise in space. *Notes:* 1. This definition is consistent with observing a clockwise rotation when the electric field vector is viewed in the direction of propagation. 2. A right-handed helical antenna radiates a right-hand polarized wave.

It is important to note that the above definition of RCP was introduced in 1942 when the IEEE was still known as the Institute of Radio Engineers (IRE), so any radio astronomer not using the proper definition has little excuse! As we've pointed out, this definition is opposite to that of physicists and optical astronomers. Obviously, the sense of the circular polarization will be reversed if one views the electric vector *along* the direction of propagation from the perspective of the emitter or source, rather than *against* the direction of propagation from the perspective of the observer or radio telescope; therefore, care must be taken to keep the IEEE definition within reach when calibrating circular polarization. If one does not have a copy nearby, the following mnemonic will help the radio polarimetrist get by: first, stick out your hands with open palms and point your thumbs in the direction of propagation of the incoming radio waves. Now, slowly make a fist. Whichever hand has its fingers curling in the directon of the electric field rotation defines the handedness of the circular polarization.

Once one has carefully defined the sense of circular polarization, one is only halfway to the promised land of absolute circular polarization calibration. What remains is to properly define the Stokes parameter V. Is it the difference between LCP and RCP, or the difference between RCP and LCP? And whose definition of RCP and LCP should be used? These are weighty matters, so they are best left up to the International Astronomical Union (IAU), and luckily these matters have been addressed by this supranational governing body. There is good news and there is bad news. The good news first: in 1973, the IAU's Commission 40 (Radio Astronomy) met in Sydney, Australia to establish a resolution that provides an absolute definition of the Stokes V circular polarization. The bad news is that they don't seem to have told anybody about it. The author struggled to find the IAU proceedings in which this definition is established; adding to the difficulty, the IAU does not provide the definition anywhere on the internet. The definition is so well hidden that it warrants an explicit reproduction here. From *Transactions of the IAU* (IAU 1974, p. 166):

8. POLARIZATION DEFINITIONS

A working Group chaired by Westerhout was convened to discuss the definition of polarization brightness temperatures used in the description of the polarized extended objects and the galactic background. The following resolution was adopted by Commissions 25 and 40: 'RESOLVED, that the frame of reference for the Stokes parameters is that of Right Ascension and Declination with the position angle of electric-vector maximum, θ, starting from North and increasing through East. Elliptical polarization is defined in conformity with the definitions of the Institute of Electrical and Electronics Engineers (IEEE Standard 211, 1969). This means that the polarization of incoming radiation, for which the position angle, θ, of the electric vector, measured at a fixed point in space, increases with time, is described as right-handed and positive.'

This unambiguously states that if one uses the radio definitions for RCP and LCP, then the Stokes V parameter for circular polarization is defined as the difference between RCP and LCP:

$$\boxed{V \equiv \text{RCP} - \text{LCP}.}$$

(2.56)

Therefore, radiation with pure RCP would be observed to have positive Stokes V, and a negative intensity would be seen in a Stokes V spectrum for radiation with pure LCP.

In an astronomical context, the magnetic field direction is defined by convention as positive when pointing away from the observer and negative when pointing toward the observer; this convention also applies to astronomical velocities. If the sense of circular polarization has been defined according to the IEEE convention, and Stokes V has been defined according to the IAU convention, then a positive magnetic field will produce Zeeman splitting in an *emission* line such that the RCP σ component will be shifted to lower frequency and the LCP σ component will be shifted to higher frequency. Therefore, the difference will yield a Stokes V profile that is positive at frequencies lower than the line center and negative at frequencies above the line center. Table 2.2 shows the expected Stokes V profiles for different scenarios given the proper radio astronomical conventions.

Ideally, all astronomers should be using the same definition of Stokes V. Since optical astronomers have the opposite definitions of RCP and LCP from radio astronomers, this suggests that an optical astronomer would then need to reverse the difference in equation (2.56). We believe, even though the IAU resolution for the definition of Stokes V is phrased in terms of the radio convention for circular polarization, and was created by the Radio Astronomy working group, that the resolution should be followed by *all* astronomers. However, before enforcing a definition across all wavelengths, radio astronomers would do well to get on the same page. A thorough analysis of Zeeman observations over the last 40 years has revealed a disturbing disregard for the conventions of circular polarization, which is somewhat surprising given that the IAU resolution was made 35 years ago.

2.4.2.1   A Brief History of Confusion

As we have seen, a magnetic field sign reversal can easily be achieved when comparing observations of the same field by different authors: all it takes is one differing convention. In practice, all authors should state their conventions carefully in radio (or any) polarization work. We have compiled the conventions used in previous single-dish[30] circular polarization radio observations (regardless of whether a Zeeman detection was made) in order to collate which, if any, circular polarization and Stokes V conventions were explicitly or implicitly used. Obviously, in order to compare magnetic field directions for a given source observed by different investigators, possibly using different telescopes, these conventions must be stated.

The results are presented in Table 2.4, with column (1) showing the citation of the paper or textbook, column (2) indicating whether the authors used the IEEE (or IRE) convention, and column (3) listing the Stokes V definition employed by the authors (if a mathematical formulation is presented, the original notation is reproduced; if only a statement of the Stokes V definition is made, we infer the symbolic representation, using RCP and LCP, and make a note that the formulation was inferred). Column (4) contains notes indicating how the Stokes V definition was

---

[30]We have included a small set of interferometric observations when they are relevant to single-dish observations; Hamaker & Bregman (1996) point out that all interferometric polarization work has used a 'black-box' formula derived by Morris et al. (1964). This formulation, devised before the IAU convention, is for $V = \text{LCP} - \text{RCP}$ (Thompson et al. 2001). It is not straightforward to parse this from the stated complex Stokes visibility relations.

inferred. Column (5) contains relevant excerpts from the paper or our own commentary regarding the conventions. We include a selection of radiation, radio astronomy, and polarimetry textbooks as well.[31] This is not intended to be a complete sample of the historical literature, rather a fairly uniform (and hopefully not-too-biased) sampling of the observational record. We note that Table 2.2 was invaluable in assessing the definition of Stokes V from plots in papers where the definition was either not stated at all or was stated ambiguously.[32]

It is immediately obvious from Table 2.4 that the convention for the sense of circular polarization is explicitly stated less than half of the time. This is somewhat disconcerting. The good news is that 100% of the time when it does appear, the IEEE convention is used! It therefore seems to be universally true that radio astronomers have adopted the IEEE definitions of RCP and LCP. We reiterate that this is very much expected since the convention was introduced in 1942 and has been presented in every radio astronomy textbook, starting with Pawsey & Bracewell (1955). This is not to say that those radio astronomers who did not list a convention used the proper one, but we adopt the generous assumption that all radio circular polarization work has been conducted using the proper IEEE definitions of RCP and LCP.

The real problem is the definition of Stokes V. The Stokes parameters were introduced to radio astronomy in 1958 by Cohen (1958), who, while using the IEEE conventions for LCP and RCP, defined Stokes V as left-handed minus right-handed circular polarization. Then Kraus (1966), arguably the most common reference text for radio astronomers,[33] followed suit. Historically, this is not an error, as the IAU definition was established in 1973 and the definition was completely arbitrary before this time. But if Kraus (1966), which is within arm's length of almost every radio astronomer, were the only reference one used when defining Stokes V, one can see that it would be very easy to avoid using the proper 1973 IAU definition. This might be why so many of the authors in Table 2.4 use $V = \text{LCP} - \text{RCP}$ right up to the present, even though Kraus (1966) is rarely ever cited when presenting conventions. It can also be seen that a handful of authors were consistently using the non-Kraus IAU convention before it was resolved into existence. And the most disturbing trend is that the Stokes V definition changes for any given primary author with time!

---

[31] No optics texts are included since we have noted that every optics textbook, including the canonical Born & Wolf (1999), uses the non-IEEE conventions.

[32] Table 2.2 can be used to infer the field direction for a Stokes V spectrum that was defined using LCP − RCP by simply finding the matching S curve shape for the corresponding measurement scenario (i.e., the plotted V profile is a function of frequency or velocity, and corresponds to a Stokes I profile for an emission or absorption feature) and associating the field direction opposite to that listed in the table.

[33] Often referred to in the field as "The Bible."

TABLE 2.4

CIRCULAR POLARIZATION AND STOKES V CONVENTIONS IN RADIO OBSERVING

Reference (1)	IEEE (2)	Stokes V (3)	IAU (4)	Note (5)	Excerpts & Comments (6)
Textbooks					
Pawsey & Bracewell (1955)...	Yes	$\cdots$	$\cdots$	$\cdots$	Cite IRE definition.
van de Hulst (1957)	No	$a^2 \sin 2\beta$	No	1	For classical LCP, $\tan\beta = -1$ and $V < 0$.
Piddington (1961)............	Yes	$\cdots$	$\cdots$	$\cdots$	Only a description of handedness.
Steinberg & Lequeux (1963)..	$\cdots$	$\cdots$	$\cdots$	$\cdots$	
Kraus (1966)	Yes	$S_L - S_R$	No	1	Cites IRE definition.
Christiansen & Högbom (1969)	$\cdots$	$B_{\rm rc} - B_{\rm lc}$	Yes	1	Unable to find any convention for RCP or LCP.
Rybicki & Lightman (1979))..	No	$RCP - LCP^{\rm a}$	No	1	Use physics convention for handedness; define $V > 0$ for non-IEEE RCP.
Shu (1991)	$\cdots$	$\cdots$	$\cdots$	$\cdots$	Refrains from taking a stance: "Considerable room for confusion on conventions exists here; so if you need to measure circular polarization, always check on the sign conventions being used by other people."
Stutzman (1993)	Yes	LCP − RCP	No	2	Follows (Kraus 1966).
Rohlfs & Wilson (1996)	Yes	RCP − LCP	Yes	2	Description of circular polarization matches IEEE; statement of Stokes V equivalent to IAU.
Tinbergen (1996)............	Yes	$I_{\rm rc} - I_{\rm lc}$	Yes	1	Does no favors by dancing around V issue: "Given the apparently contrary conventions of radio and 'traditional' optical astronomers, I hesitate to recommend either." Points out an IAU convention exists. In Fig. 2.1 caption, adopts IEEE and IAU conventions for circular. States elliptical polarization Jones vector using IEEE convention on p. 59., then on p. 61 defines $V = I_{\rm rc} - I_{\rm lc}$.
Burke & Graham-Smith (1997)	Yes	$RCP - LCP$	Yes	1	Explicitly state both IEEE and IAU conventions.
Jackson (1998)..............	No	$a_+^2 - a_-^2$	Yes	1	Here a_+^2 corresponds to *positive helicity*, which is LCP in the classical physics convention.
Thompson et al. (2001)	Yes	RCP − LCP	Yes	2	Explicitly state both IEEE and IAU conventions.
Articles					
Cohen (1958)...............	Yes	$I_l - I_r$	No	1	Cites IRE definition.
Morris et al. (1963)	Yes	L.H. − R.H.	No	1	Cite Pawsey & Bracewell (1955) and radio convention in defining L.H. and R.H.
Barrett & Rogers (1966)	Yes	LCP − RCP	No	1	Cite IEEE definition, but not Stokes V sense; results match Raimond & Eliasson (1969), who used IEEE and defined $V = $ LCP − RCP.
Coles et al. (1968)............	Yes	$T(\mathrm{R}) - T(\mathrm{L})$	Yes	1	"$T(\mathrm{R})$ and $T(\mathrm{L})$ are the temperatures for right- and left-circular polarization, defined with respect to the direction of propagation."
Ball & Meeks (1968).........	Yes	$T(R) - T(L)$	Yes	1	"We consider right circular polarization as rotation of the electric vector in a clockwise sense when viewed along the direction of propagation. This is the standard radio definition, which is opposite to the definition used in optics."
Verschuur (1968).............	$\cdots$	RCP − LCP	$\cdots$	2	States that "difference between right- and left-hand polarization incident on the feed" is plotted in text and Fig. 1 caption. Incorrectly infers negative fields in Cas A absorption: should be positive.
Davies et al. (1968)	Yes	L − R	No	1	Caption of Fig. 1 states "polarization is plotted as left minus right hand (IRE definition) incident on the telescope." Also, frequent calibration was performed by radiating a circularly polarized CW signal into the horn. Derive a positive field for Cas A.

TABLE 2.4—*Continued*

Reference (1)	IEEE (2)	Stokes V (3)	IAU (4)	Note (5)	Excerpts & Comments (6)
Verschuur (1969a)	$\cdots$	RCP $-$ LCP	Yes	2,3,4	States Zeeman spectrum is "difference between right- and left-hand polarization." Later formulates spectrum as $LH_j - RH_j$. Then in Fig. 3 caption says V is "right-hand minus left-hand polarization, incident on the dish."
Verschuur (1969b)	$\cdots$	RCP $-$ LCP	Yes	2,3,4	V plotted as "right-minus-left hand polarization incident on the telescope."
Raimond & Eliasson (1969)..	Yes	$I_{\text{left}} - I_{\text{right}}$	No	1	No explicit IEEE mention, but "Right-hand polarization is defined as a clockwise rotation of the electric vector when viewed along the direction of propagation."
Coles & Rumsey (1970)	Yes	$T(R) - T(L)$	Yes	1,3	1665 MHz W49 V spectrum matches that of Ball & Meeks (1968).
Brooks et al. (1971)	$\cdots$	RCP $-$ LCP	Yes	2,3,4	Used Parkes to meaure "right-hand minus left-hand polarization profile" toward Ori A; observe the same profile as Verschuur (1969b).
Weiler (1973)	Yes	RCP $-$ LCP	Yes	2	Interferometric treatment of WSRT polarization: "$+V$ corresponds to Right Hand Circular Polarization (IRE Definition)."
Conway (1974)	Yes	RCP $-$ LCP	Yes	2	Points out the "deplorable confusion" surrounding sign and terminology of circular polarization work.
IAU (1974).................	Yes	RCP $-$ LCP	Yes	2	The original IAU definition: "right-handed" circular = "positive" V.
Crutcher et al. (1975)........	$\cdots$	RCP $-$ LCP	Yes	2	The "Zeeman profile" is "difference between right and left circularly polarized power spectra."
Verschuur (1979)	Yes	$I_{\text{RH}} - I_{\text{LH}}$	Yes	1	Computed the "Stokes parameter V spectrum."
Crutcher et al. (1981)........	$\cdots$	$\cdots$	$\cdots$	$\cdots$	No definition. Can't be inferred from field signs because no detections were made, only upper limits.
Troland & Heiles (1982a)....	$\cdots$	LCP $-$ RCP	No	2,4	Only mention the "Stokes parameter V spectrum" is the "difference between the line profiles detected in opposite senses of circular polarization." For inferred positive fields in emission lines, V must have been defined as LCP $-$ RCP(i).
Heiles & Troland (1982).....	$\cdots$	RCP $-$ LCP	Yes	2,4	State V is the "difference between right-hand and left-hand circular polarization." Derived positive fields for both emission and absorption consistent with right minus left.
Bregman et al. (1983)	$\cdots$	LCP $-$ RCP	No	2,3,4	Interferometric Cas A observations show V matching Davies et al. (1968) and Heiles & Troland (2004), therefore $V = $ LCP $-$ RCP. Curious because they use WSRT and cite Weiler (1973), who define RCP $-$ LCP, in calibration discussion.
Crutcher & Kazes (1983)....	$\cdots$	LCP $-$ RCP	No	2,3	V is "the difference between the left and right polarization spectra." If truly LCP $-$ RCP, derived fields should be *positive* for OH absorption in Ori B. Incorrectly infer a negative field.
Heiles & Stevens (1986).....	$\cdots$	RHC $-$ LHC	Yes	1	Only mentioned in figure captions. Measure positive field for Ori B, opposite to Crutcher & Kazes (1983).

TABLE 2.4—*Continued*

Reference (1)	IEEE (2)	Stokes V (3)	IAU (4)	Note (5)	Excerpts & Comments (6)
Troland et al. (1986)	⋯	LHC − RHC	No	1,4	V sense only listed on plot. Negative field in Ori A OH absorption consistent with 21 cm Zeeman and consistent with $V = \text{LHC} - \text{RHC}$. However, they claim they "obtained an H I Zeeman effect spectrum with the Nançay telescope" and that "it is similar to" those of Verschuur (1969b) and Brooks et al. (1971): however, the latter two papers show RHC − LHC and have Stokes V flipped in sense from this result.
Kazes & Crutcher (1986)	⋯	RCP − LCP	Yes	4	No V definition; must be RCP − LCP for the inferred negative fields in W22 absorption.
Schwarz et al. (1986)	⋯	LCP − RCP	No	3,4	Interferometric Cas A observations show V matching Davies et al. (1968), Bregman et al. (1983), and Heiles & Troland (2004), therefore $V = \text{LCP} - \text{RCP}$.
Crutcher et al. (1987)	⋯	RCP − LCP	Yes	2	Define V as "right minus left circularly polarized power."
Heiles (1987)	Yes	RCP − LCP	Yes	2	"There is confusion with regard to signs." Points out IEEE convention. "In Zeeman splitting, if the RHC component is observed at a higher frequency then the magnetic field points towards the observer." This is true for emission lines if $V = \text{RCP} - \text{LCP}$.
Heiles (1988)	⋯	LCP − RCP	No	4	Only states "V is the difference between two circular polarizations." Positive field in emission in Fig. 2 consistent with $V = \text{LCP} - \text{RCP}$.
Kazes et al. (1988)	⋯	RCP − LCP	Yes	4	No V definition; for derived positive field in S106 absorption, must have plotted RCP−LCP.
Heiles (1989)	Yes	LCP − RCP	No	1	"V is the difference between the two circular polarizations, equal to left-hand *minus* right-hand circular (LCP-RCP, in the IEEE definition, in which RCP is defined as a clockwise rotation of the electric vector as seen from the transmitter of the radiation, Kraus 1966). This convention has also been used in several of our previous publications (Troland & Heiles 1982b; Heiles & Troland 1982; Heiles 1988)." This is not true of Heiles & Troland (1982).
Goodman et al. (1989)	Yes	RCP − LCP	Yes	2,4	"The Stokes V spectrum is a display of the right minus left cicularly polarized signals. The sense of our left and right circular polarizations follows the IEEE definition; that is, right circular polarization is clockwise rotation of the electric vector as seen from the transmitter of radiation."
Fiebig & Güesten (1989)	⋯	$T_\text{B}(\text{RHC})$ $- T_\text{B}(\text{LHC})$	Yes	1	Cite Heiles (1987) for definition of V; luckily for them, this is one of the places Heiles states the proper IAU definition.
Güsten & Fiebig (1990)	⋯	RCP − LCP	Yes	2	V is "difference between the right- and left-handed circularly polarized emission."
Kazes et al. (1991)	⋯	LCP − RCP	No	4	LCP − RCP is inferred from derived field signs.
Heiles et al. (1993)	⋯	RCP − LCP	Yes	2	The Stokes V spectrum is observed "by subtracting the left from the right circular polarization."
Crutcher et al. (1993)	⋯	$T_r - T_l$	Yes	1	
Davies (1994)	⋯	LCP − RCP	No	2,3,4	$B_\parallel$ obtained from "left minus right polarization."
Goodman & Heiles (1994) ...	⋯	RCP − LCP	Yes	1,4	
Güsten et al. (1994)	⋯	LHC − RHC	No	1	
Verschuur (1995a)	⋯	RCP − LCP	Yes	2	V plots show "right-hand−left-hand circular polarization incident on the telescope."

TABLE 2.4—*Continued*

Reference (1)	IEEE (2)	Stokes V (3)	IAU (4)	Note (5)	Excerpts & Comments (6)
Verschuur (1995b)	· · ·	$RCP - LCP$	Yes	2	V displayed "as right-hand minus left-hand circular polarization."
Hamaker & Bregman (1996) .	Yes	$RCP - LCP$	Yes	2	An authoritative account of IEEE and IAU circular conventions and their use in radio interferometic polarization. Find disturbing result: all interferometric "radio polarimetry work has until now been based on a formula published first by (Morris et al. 1964)." This complex 'black-box' formula, when decoded, yields $V = LCP - RCP$. Thompson et al. (2001) points out that (Morris et al. 1964) "predates the IAU definition and follows an earlier convention." (The 1986 edition claimed they were in agreement with (Morris et al. 1964).) Weiler (1973) rederived the black-box formula and found that it did not comply with the IAU convention.
Heiles (1996)	Yes	$LCP - RCP$	No	1	"In the present and our previous papers, we follow the definition of Stokes parameters given by Kraus (1966), to wit: (1) We use the IEEE definition, in which left circular polarization (LCP) rotates clockwise as seen by the receiving antenna; and (2) $V = LCP - RCP$."
Troland et al. (1996)	· · ·	$R - L$	Yes	1,4	Only mention $V = R - L$ in figure captions. Mention sign of ρ Oph detection incorrectly reported as positive in Crutcher et al. (1993).
Crutcher et al. (1996)	· · ·	$T_{\rm R} - T_{\rm L}$	· · ·	1	State that $V \equiv T_{\rm R} - T_{\rm L}$, but inferred field directions from Stokes V spectra are either incorrect or the Stokes V spectra are truly plotted as $T_{\rm L} - T_{\rm R}$.
Crutcher et al. (1999)	· · ·	$L - R$	No	1	State that "middle panels show the observed $V = L - R$ spectrum." However, inferred negative fields for emission features only possible if $V = R - L$.
Crutcher & Troland (2000) . . .	· · ·	$LCP - RCP$	No	4	V is not defined, but a positive field is inferred for emission in L1544; V must be plotted as $LCP - RCP$. Astronomical and ground-based circular polarization calibration was performed.
Uchida et al. (2001)	· · ·	$LCP - RCP$	No	2	"Stokes-V spectrum (the difference between the left- and right-circularly polarized spectra)."
Heiles (2001)	Yes	$i_{\rm LCP} - i_{\rm RCP}$	No	1	Explicitly follows IEEE convention for LCP. States that Tinbergen (1996) defines V "opposite ours and, also, the conventional one used by radio astronomers." Former is true, latter false.
Bourke et al. (2001)	· · ·	$I_{\rm RCP} - I_{\rm LCP}$	Yes	1	Inferred positive fields Ori B absorption consistent with $RCP - LCP$.
Heiles (2002)	· · ·	$E_{LCP}^2 - E_{RCP}^2$	No	1	Incorrectly labels Stokes V as "the IEEE definition."
Heiles & Troland (2004)	· · ·	$LCP - RCP$	No	2,3,4	V formed "by subtracting RCP from LCP." Checked polarization sense with a calibration helix.

Notes.—IEEE RCP and LCP conventions are generously assumed for all papers for which no circular polarization convention was stated. (1) Stokes V explicitly defined mathematically; (2) Stokes V definition and conformity with IAU convention are inferred from text, no formulation given; (3) Stokes V definition is inferred from plot of Stokes V and its comparison to the Stokes V plot of an author who has defined Stokes V; (4) Stokes V definition is inferred from plot of Stokes V and the derived sign of a detected magnetic field.
[a] Here, classical RCP is equivalent to the IEEE LCP.

As an example of how confusing these conventions can be, we follow the trail of the published Stokes V spectrum toward Cas A. Its very first appearance is the Green Bank 140 ft result, plotted in Fig. 1 of Verschuur (1968). The caption of this plot clearly states that Stokes V is defined as RCP − LCP, with no convention defined for RCP or LCP. The circular polarization features that are associated with the Perseus arm absorption features (near LSR frequency +200 kHz or LSR velocity −45 km s^{-1}) are then clearly associated with a positive field, as can be inferred by comparison with Table 2.2. However, in the text, the reported fields are negative. Three months later, Davies et al. (1968) confirmed the detection using the Jodrell Bank Mk II 125 × 85 ft, but clearly plot (in their terminology) L − R. Here again no convention for LCP or RCP is defined, but we assume the IEEE/IRE definition is in use. The resulting Stokes V pattern shown in their Fig. 2 is inverted when compared to that of Verschuur (1968). However, they properly derive a *positive* field. Verschuur (1969a) thoroughly obfuscates the situation: he continues to present RCP − LCP in his Fig. 3, clearly stating that his V spectrum is "right-hand minus left-hand polarization, incident on the feed" (in both the text *and* the figure caption), but then proceeds to define his derived spectrum as $LH_j - RH_j$ divided by a "band pass." In hindsight, we recognize that this was a typographical error, and that he did indeed plot RCP − LCP. Verschuur (1969a) continues his discussion of Cas A with a note added in proof: "The signs given on [*sic*] Paper I were incorrect." A month later, Verschuur (1969b) reports new Cas A observations and notes that his field detections "have also been independently observed (Davies et al. 1968), but with an opposite field direction. I have checked this calibration in the new experiment (December 19, 1968, to January 3, 1969) and find that the field direction was in error." It is strange that the calibration is mentioned as a reason for the error: his RCP − LCP spectrum is displayed properly and his circular polarization sense is correct. He simply did not infer the correct sense of circular polarization for the shifted Zeeman components in absorption. Ten years later, Verschuur (1979) expands his explanation slightly:

> Due to a unique error in the calibration of the sense of polarization of the output data, a change in the sense of the magnetic fields detected was introduced into the system. Thus the first paper on these positive results (Verschuur 1968) had an incorrect sign in the field vector.

Again, the sense of polarization for the output data is properly calibrated; it remains a mystery why Verschuur continued to claim otherwise. Recently, Heiles & Troland (2004) presented carefully calibrated Stokes V spectra for the 21 cm absorption features toward Tau A (using Arecibo) and Cas A (using archival data from the Hat Creek 85 ft). Their Stokes V spectra for both sources, which are explicitly plotted as LCP−RCP, are inverted from those presented in Verschuur (1969a).

They claim that Verschuur (1969a) "stated that his Stokes V was IEEE LHC − RHC, but this appears to be nothing more than a typographical error. It is not a fundamental sign error because the signs of his derived $B_\parallel$ are correct." This statement is not entirely true:[34] we have seen that both Verschuur (1968) and Verschuur (1969a) state that RCP − LCP is being plotted, and in neither case is the IEEE convention ever invoked. However, since Davies et al. (1968) do use the IEEE/IRE convention, and their LCP − RCP spectrum is identical to that of Heiles & Troland (2004), and is the inverse of Verschuur's RCP − LCP, we can safely infer that Verschuur (1969a) did use the IEEE convention for RCP. The typographical error does indeed exist in the section labeled *Display of Data*, but Verschuur (1969a) explicitly says here, and in the caption, that he is plotting RCP − LCP. Happily, both Verschuur (1969a) and Heiles & Troland (2004) agree on the signs of the fields regardless of the confusion over the Stokes V definition. More confusion was introduced in 1983 by interferometric observations of Cas A using the Westerbork Synthesis Radio Telescope (Bregman et al. 1983). These results, later confirmed by Schwarz et al. (1986), showed a Stokes V profile that matches Davies et al. (1968) and Heiles & Troland (2004). The intricacies of interferometric polarization calibration are beyond our scope (for an authoritative discussion, see Hamaker & Bregman 1996); but given that the inferred field is positive, it is clear that the resulting calibrated spectrum for this experiment yielded the non-IAU compatible $V = \text{LCP} − \text{RCP}$. Davies (1994) decided to revisit the Cas A Zeeman experiment using the Mk IA 250 ft telescope at Jodrell Bank. They used a new autocorrelation spectrometer along with the same feed and receiver system as those on the Mk II (with which the original 1968 confirmation was made). The same Stokes V profile was observed, defined again as $V = \text{LCP} − \text{RCP}$. Finally, we note that we have observed Cas A using the GBT 100m, and after careful calibration of our Stokes V spectrum, for which we *of course* used the IAU definition of RCP − LCP, we (thankfully) infer a positive field direction in the Perseus arm absorption features.

The case of Cas A highlights three important things. First, the proper IAU Stokes V definition is unlikely to be unanimously adopted in the future; and even if it is, when consulting the historical record, one must pay careful attention to the Stokes V convention that was in use for the presented results. Next, the Zeeman splitting of Cas A was reproducible at six of the major single-dish radio telescopes across the world (and possibly more that the author is not aware of): if the Stokes V spectrum is calibrated correctly, then there should be no disagreement between observers using different telescopes or different conventions in the inferred direction of the line-of-sight magnetic

---

[34]The confusion is perpetuated by Heiles & Crutcher (2005), who state "Verschuur made a typographical error in labeling the signs of his Stokes V profiles."

field. Finally, even if calibration is perfectly achieved, there is room for *interpretive error* on the part of the observers: they must correctly attribute the RCP (LCP) component to the higher-frequency Zeeman splitting component in absorption (emission) for positive fields. It can be seen that a number of authors in Table 2.4 incorrectly inferred the *sign* of the field from properly displayed Stokes V spectra, including the original detection by Verschuur (1968). One might note that Zeeman's own detection in 1896 was plagued by the observer's misinterpretation of circular polarization sense; we suggest that the folly of interpreting the incorrect field direction from a carefully and correctly calibrated Stokes V spectrum be named "Zeeman's curse."[35]

Since calibration of the circular polarization is the key to a confident determination of magnetic field direction via the Zeeman effect, we turn our attention to these details.

2.4.2.2   One More Thing To Keep Track Of: Reflections

The sense of circular polarization flips when radiation is reflected[36] (see § 8.2 of Stutzman 1993 for a thorough description). Receivers located at the prime focus of a telescope (e.g., the Green Bank 140 ft, the Hat Creek 85 ft,[37] the GBT for $\nu < 1230$ MHz, the Effelsberg 100 m for more than half of their receivers, the Parkes 64 m, the Lovell 250 ft, and the Jodrell Bank Mk II 125 × 85 ft) will have RCP and LCP reversed from their astronomical sense, because the incoming radiation undergoes a single reflection from the primary antenna; the received Stokes V spectrum will therefore be reversed in sign from its true astronomical sense. Receivers at the secondary focus of a telescope (e.g., the GBT for L band and higher frequency receivers, Effelsberg for some receivers) will be receiving the true astronomical sense of the circular polarizations, because the incoming astronomical signal is reflected twice: once from the primary and once from the secondary reflector; therefore the receiver sees a true astronomical Stokes V spectrum. Finally, since Arecibo has a tertiary reflector, and all receivers are located at the tertiary focus, the circular polarizations for all Arecibo measurements are reversed from their astronomical sense: the receivers at Arecibo always see a Stokes V spectrum that is flipped in sign from its astronomical sense.

---

[35]Zeeman had the luxury of correcting his error before the English translations of his work were published.

[36]This is not true for large angles of incidence.

[37]These first two telescopes are no longer operational, but plenty of Zeeman results were produced by both.

2.4.2.3 Getting It Straight with Calibration

Since the Zeeman effect allows us to probe the direction of the line-of-sight magnetic field as well as the strength, one ought to take care to calibrate the circular polarization so that the correct sign can be attributed to the field direction. We have seen that, even if the correct IEEE and IAU conventions are stated for the incoming radiation, an odd number of reflections at the telescope can flip the circular sense measured at the receiver; if not taken account of, the opposite field direction will be inferred. There are two ways to calibrate the Stokes V spectrum. The first is preferable, the second is the easiest.

First, one can transmit circularly polarized radiation of a known sense into the telescope beam and measure the resulting Stokes V signal. At Arecibo, this has been done many times using helical antennas (a rear-fed conical right-handed helix transmits RCP); the transmitter is located on the ground, so the radiation is transmitted directly onto the secondary through an opening in the 305 m primary. Since two reflections occur, the feed will receive the sense of circular polarization that is transmitted. Crutcher & Troland (2000) performed these ground-based calibrations by transmitting RCP and LCP signals into the Gregorian dome. We performed the same calibration at the GBT. A front-fed conical right-handed helix (which, unlike a rear-fed RH helix, transmits LCP) was used to directly transmit LCP into the L-, C-, and X-band receivers. The resulting Stokes V spectrum indicated an LCP signal was being received. Each of these receivers is located at a secondary focus on the feed arm such that an astronomical signal will undergo two reflections; an astronomical LCP signal will be measured as an LCP signal at the receiver.

A secondary calibration method is to observe a source of known circular polarization. The Galactic mainline OH masers are an obvious choice for circular polarization calibration: because of their large fractional circular polarization, as well as their extreme brightness, these targets require only a very short integration to completely characterize the sense of circular polarization. Obviously, there is a danger involved because this method of calibration bootstraps one's measured Stokes V spectrum to that of some previous observer. Therefore, one must trust that the signs and conventions for circular polarization were properly established for the historical observations. For the purpose of single-dish circular polarization calibration, the most reliable calibrators are those of Ball & Meeks (1968) and Coles & Rumsey (1970), who all used the proper IEEE and IAU conventions (years before the IAU convention existed) and who are all known as careful and meticulous observers (W. J. Welch 2008, private communication). Observations of the absorption lines in Cas A or Ori A at 21 cm, or Ori B at 1665 or 1667 MHz, can also be used, but beware of the prevalent historical tendency of authors to plot the Stokes V spectrum for each of these sources as LCP − RCP: regardless of Stokes V convention, the derived fields for each should be positive, negative, and positive, respectively. Using the GBT, we have measured the Zeeman splitting of the 21 cm absorption in Cas A and Tau A, and of OH maser emission in W49 and W3. The results are consistent with our ground-based calibration as well as the historical astronomical observations.

2.4.3   Cross-Correlation Spectropolarimetry

Thus far we have only considered the path of the radiation from the source to the feed. What happens after polarized radiation arrives at the feed? The signal that is received by the two probes of a dual orthogonal feed will be amplified, be mixed to various frequency stages, travel from the receiver to the correlator room on optical fibers of different lengths, and fed to a correlator. This last step is where the signal is turned into a time-averaged spectrum and digitized.

A correlator can be built to produce spectra in two ways (Heiles 2001; Thompson et al. 2001). In the first, time-averaged correlation functions are calculated, and then the Fourier transform is performed on the correlation functions. This is known as an XF, or lag, correlator; the GBT Autocorrelation Spectrometer is an XF correlator. In an FX correlator, the Fourier transform is performed on the signals from the two feeds, then the autocorrelation and cross-correlation spectra are created by multiplying each signal with the complex conjugate of another. The GBT Spectral Processor is an FX correlator.

Usually, astronomers aren't interested in the cross-power spectra and therefore run a correlator in autocorrelation mode. The output will be the autocorrelation of the two orthogonal feeds. However, one can run a correlator in a mode that produces cross-products in addition to auto-products; we call this *full-Stokes mode*, because the entire suite of Stokes parameters can be built from the correlator output. Without presenting the details, we state without proof that in full-Stokes mode, the correlator outputs the autocorrelation spectra (call these AA and BB) and the real and imaginary parts of the complex cross-correlation power spectrum (call this AB; see Robishaw & Heiles 2006 for details). We can then write the Stokes parameters in terms of the correlator output:

$$
\begin{aligned}
I &\equiv AA + BB \,, \\
Q &\equiv AA - BB \,, \\
U &\equiv 2\,\mathrm{Re}(AB) \,, \\
V &\equiv 2\,\mathrm{Im}(AB) \,.
\end{aligned}
\tag{2.57}
$$

The signals that are incident on the feed probes will undergo many phase and intensity changes on the way to the correlator; therefore we need a way of representing the transformation between the Stokes parameters that are incident on the feed and those that are output from the correlator. Luckily, just such a 4×4 transformation matrix exists; it is known as the Mueller matrix and we shall discuss its details in § 2.4.3.1.

If trying to measure a small linear polarization signal, one should use a native dual-circular feed (Christiansen & Högbom 1969): the difference between two large outputs from a dual-linear feed will be less sensitive to small values than the cross-correlation of two circulars. Likewise, if trying to measure a small circular polarization signal, it is far better to use a native dual-linear feed. Yet, for almost every single published Zeeman observation that has been made using single-dish telescopes with a native dual-linear feed, the observers have introduced a phase shift of $90°$ between the linear feed responses in order to generate orthogonal circular polarizations! Heiles (2002) discusses the pitfalls of using a hybrid in cross-correlation polarimetry. In particular, one is safe in doing so if the hybrid is placed before the first amplification stage. If this were the case, a native dual-circular feed could be converted to a dual-linear feed. One could then use the advantage of cross-correlation to detect weak circular polarization. If the hybrid comes after the first amplifier, any advantage is lost. On the GBT, there is no pre-amplification hybrid option for any receiver. This is most likely because the hybrid would need to be cooled: since it is in front of the first amplifier, the noise introduced by the hybrid is amplified by the same amount as the incoming signal. One can only surmise that it is the fear of implementing the details of cross-correlation spectropolarimetry, and the ease of throwing the hybrid switch, that leads to the unfortunate trend of converting native dual-linear feeds to dual-circular by means of a hybrid.

2.4.3.1   The Mueller Matrix

> The measurements have been corrected for everything you correct such measurements for.
>
> GART WESTERHOUT, 1966

In 1943, a physics professor at the Massachusetts Institute of Technology named Hans Mueller introduced a method of transforming Stokes parameters in the form of 4×4 matrices (Mueller 1948).[38] These have since become known as *Mueller matrices*. This optical calulus was originally developed by Soleillet (1929) and Perrin (1942), but was set in the framework of matrices by Mueller.

---

[38] It appears Mueller first presented this formalism in informal class notes for course 8.26 at MIT in the 1945–1946 school year, but its true unveiling was on 1943 November 15, in a now declassified report to the Office of Scientific Research and Development entitled *Memorandum on the polarization optics of the photo-elastic shutter* by Hans Mueller (Jones 1947).

The signal path from one feed probe to the correlator output can be thought of as a linear series of devices that operate on the input signal by changing its gain and phase. If the incoming radiation were completely polarized, we could represent the incoming radiation as a Jones vector; then we could use 2×2 Jones matrices to represent how each component alters the intensity and phase of the incoming polarized signal. A grand Jones matrix for the system could be constructed by taking the product of all of these Jones matrices; these matrix products are not commutative, so care must be taken to multiply each in the correct order. The Jones vector of the signal leaving the correlator would then be the product of the final Jones matrix and the initial Jones vector. Every one of these devices and Jones matrices would have a corresponding Mueller matrix, which describes how the device alters the Stokes vector. The total Mueller matrix for the entire system would be the product of the individual matrices. The Mueller and Jones calculi are equivalent for systems that do not depolarize; i.e., that convert completely polarized incident light into outgoing completely polarized light. However, as we saw in § 2.2.1, incoming astronomical radiation is never completely polarized and we must adopt the Mueller calculus.

Heiles et al. (2001b) provide an extremely detailed description of how one parametrizes and measures the Mueller matrix for a single-dish radio telescope. We list here their grand Mueller matrix:

$$
M_{\text{tot}} =
\begin{bmatrix}
1 & \left(\begin{array}{c} -2\epsilon \sin\phi \sin 2\alpha \\ + \frac{\Delta G}{2} \cos 2\alpha \end{array} \right) & 2\epsilon \cos\phi & \left(\begin{array}{c} 2\epsilon \sin\phi \sin 2\alpha \\ + \frac{\Delta G}{2} \sin 2\alpha \end{array} \right) \\
\frac{\Delta G}{2} & \cos 2\alpha & 0 & \sin 2\alpha \\
2\epsilon \cos(\phi + \psi) & \sin 2\alpha \sin\phi & \cos\phi & -\cos 2\alpha \sin\phi \\
2\epsilon \sin(\phi + \psi) & -\sin 2\alpha \cos\psi & \sin\psi & \cos 2\alpha \cos\psi
\end{bmatrix}, \quad (2.58)
$$

where α is the amount of coupling of one polarization into the orthogonal polarization, ΔG is the gain difference between the two orthogonal polarizations, ψ is the phase difference between a linearly polarized astronomical source and the correlated calibration noise diode (known as the "correlated cal"), ϵ is the cross-coupling between the two orthogonal polarizations (nominally brought about for a dual-linear feed if the probes are not exactly 90° apart), and ϕ is the phase angle of these coupled voltages. It can be shown that there are only seven (eight minus an irrelevant phase) independent values in the Mueller matrix (Jones 1947; van de Hulst 1957, p. 44). An eagle eye will count only five parameters in equation (2.58). One of the missing parameters is the phase angle χ of the coupling of one polarization into the orthogonal one; Heiles et al. (2001b) explicitly set $\chi = 90°$. The other is the angle by which the derived position angle of the source must be rotated in order to conform with astronomical convention; they call this angle θ_{astron}, and note that

Rcvr1_2 1420 MHz BRD0 ACS 3C286 27–JAN–2006

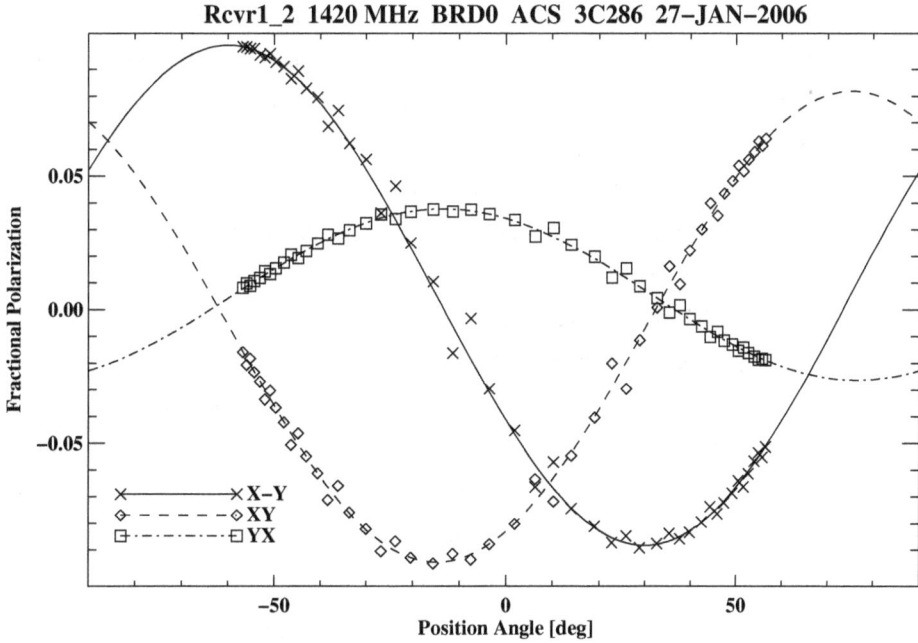

DELTAG = 0.011 ± 0.001
PSI = −19.9 ± 0.3
ALPHA = +0.4 ± 0.2
EPSILON = +0.004 ± 0.000
PHI = +158.1 ± 3.7
Q_{SRC} = −0.046 ± 0.000
U_{SRC} = −0.081 ± 0.000
POL_{SRC} = +0.094 ± 0.000
PA_{SRC} (**UNCORRECTED FOR M_{ASTRO}**) = −59.81 ± 0.12
NR GOOD POINTS: X–Y = 46 XY = 48 YX = 48 / 48

Mueller Matrix:

$$
\begin{array}{rrrr}
1.0000 & 0.0053 & -0.0079 & 0.0033 \\
0.0054 & 0.9999 & -0.0000 & 0.0133 \\
-0.0063 & -0.0045 & 0.9400 & 0.3410 \\
0.0057 & -0.0124 & -0.3411 & 0.9400
\end{array}
$$

FIG. 2.1 — Astronomical determination of the *L*-band Mueller matrix for the GBT. By tracking the strongly linearly polarized source 3C 286 over a wide range of parallactic angle, we are able to derive the relevant Mueller parameters by the method of least squares. The resulting Mueller matrix is presented at the bottom.

to obtain the *final* Mueller matrix, one must multiply the total of equation (2.58) by the Mueller matrix M_{sky}, which simply contains elements corresponding to a rotation of Q and U by θ_{astron}.

We observationally determine the Mueller matrix by tracking a linearly polarized source over a large range of parallactic angle. Figure 2.1 shows the results from an observation of 3C 286 using the dual-linear L-band receiver at the GBT. The crosses show the correlator Stokes Q output divided by the correlator Stokes I output. The diamonds show the normalized U output. Since the source is strongly linearly polarized, these two outputs should vary sinusoidally with twice the parallactic angle, and each sinusoid should have equal amplitude. It is evident that this is not quite the case. The squares represent the Stokes V response from the correlator, and Figure 2.1 reveals a major leakage of linear polarization into Stokes V. There should be no circular polarization response at all since 3C 286 is not circularly polarized! A nonlinear least-squares fit of these data yields the first seven parameters listed in Figure 2.1. These derived parameters are then used to quantify the Mueller matrix for that combination of receiver and correlator. The matrix derived for these data is listed at the bottom of Figure 2.1; the nonzero off-axis elements quantify how much one Stokes parameter leaks into another. If the correlator output is corrected by this Mueller matrix, the proper astronomical Stokes parameters incident on the feed are recovered. We demonstrate this in Figure 2.2 by applying the derived Mueller matrix to the 3C 286 data.

We have empirically determined that the Mueller matrices for various receivers at the GBT remain constant over long time scales. In practice, if cable lengths are changed ahead of the correlated cal at the receiver, the Mueller matrix will need to be redetermined. Otherwise, one should only need to assess the Mueller matrix once during an extended observing run of many weeks. The scenario is the same at Arecibo.

2.4.3.2   Intensity and Phase Calibration

The typical method for calibrating the intensity of a spectral line using a radio telescope is to inject a noise diode of known temperature into the signal path before the first amplifier. This is typically done very frequently so that any changes in the system gain can be calibrated. When calibrating polarization measurements, an extra complication is necessary: the *same* calibration signal must be injected into the signal path for each polarization. In practice, the output of a single noise diode is fed into a power splitter, then one output is fed into the signal path for one polarization and the other output is fed into the signal path for the orthogonal polarization. We refer to the diode and the splitter as the correlated cal. All of the derivable Mueller matrix parameters in equation (2.58) are related to either the feed or the correlated cal, and thus we can completely calibrate the Mueller matrix by relating the polarization of an astronomical source to the correlated cal. Once this is done, one can apply the Mueller matrix to calibrate the polarization of an incoming signal. However, a number of factors can change the relative phase of the two orthogonal polarizations

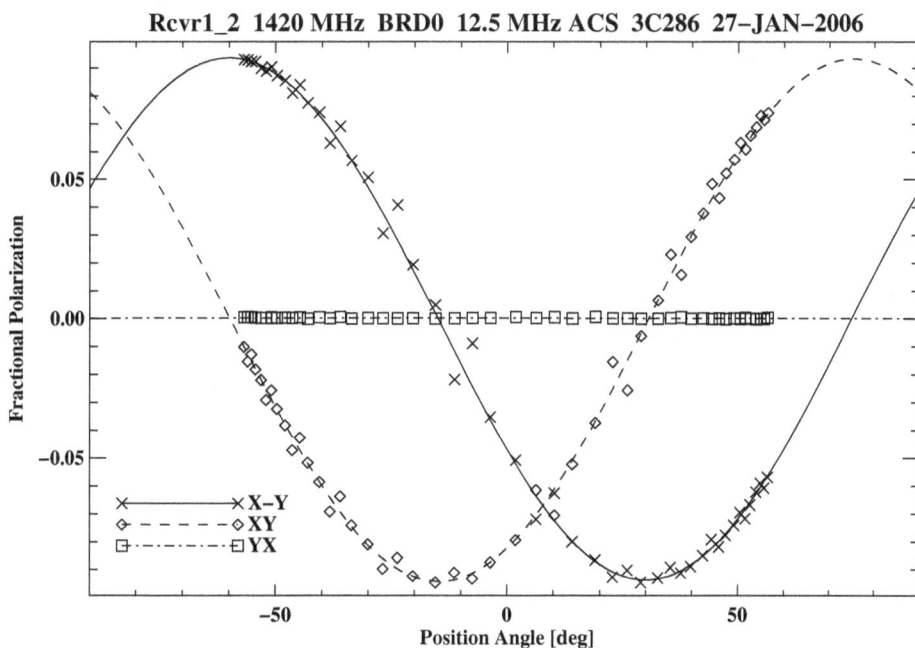

Rcvr1_2 1420 MHz BRD0 12.5 MHz ACS 3C286 27–JAN–2006

DELTAG = –0.000 ± 0.001
PSI = –0.1 ± 0.3
ALPHA = +0.0 ± 0.2
EPSILON = +0.000 ± 0.000
PHI = –21.3 ± 59.4
Q_{SRC} = –0.046 ± 0.000
U_{SRC} = –0.081 ± 0.000
POL_{SRC} = +0.094 ± 0.000
PA_{SRC} (**UNCORRECTED FOR M_{ASTRO}**) = –59.82 ± 0.12
NR GOOD POINTS: X–Y = 46   XY = 48   YX = 48 / 48

Mueller Matrix:

1.0000	–0.0001	0.0005	–0.0002
–0.0001	1.0000	–0.0000	0.0006
0.0005	–0.0000	1.0000	0.0009
–0.0002	–0.0006	–0.0009	1.0000

FIG. 2.2 — Mueller matrix applied to observations of 3C 286. The Stokes V parameter now shows no power since 3C 286 is not circularly polarized. The leakage of Stokes parameters has been minimized, as can be seen in the near-zero off-axis terms in the Mueller matrix derived from these Mueller-corrected data.

with time; therefore it is necessary to remove the phase difference frequently by injecting the correlated cal. The system contribution to the phase difference between the two polarizations can then
be removed by the methods described in Heiles (2001) and Heiles et al. (2001b).

2.4.4   Polarization Properties of Single-Dish Radio Telescopes

The spatial response of a single-dish telescope is known at the beam pattern. Most of the
radiation falls within the *main beam* of the telescope, but there are low-level responses to radiation
outside of the main beam. These responses are known as the sidelobes, with those nearest the main
beam called the near sidelobes, and not surprisingly, those far away from the main beam known
as the far sidelobes. The near sidelobe that is directly outside of the main beam is called the *first
sidelobe*; the GBT was designed with an unblocked aperture and the first sidelobe is advertised to
be 30 dB below the main beam. Heiles et al. (2001a) show maps of the Arecibo beam response
for each of the Stokes parameters. The main beam at Arecibo is elliptical, and due to the large
blockage of the primary by the support structure that hangs over the dish, the sidelobe response is
a complex function of azimuth and zenith angle, and is suppressed from the main beam response
by $\sim$12 dB.

For an alt-az telescope, the beam rotates with respect to the sky.[39] We saw in the previous
subsection that this rotation allows us to measure the variation of the polarization response for
a linearly polarized source when tracked through a wide range of parallactic angle. If one is
measuring the Zeeman effect from an emission line that is spatially extended relative to the main
beam size, the polarized beam structure can interact with the extended emission in such a way
that spurious polarized signals can be created. Since, as we shall see, the Stokes V response has
a direction on the sky, one can possibly account for this spurious emission if it shows parallactic-
angle dependence.

We now discuss the polarized beam structure of the GBT, and describe the instrumental contribution to Stokes V. But before we do this, we mention briefly the proper definition of the
parallactic angle; it would seem that this should be a well-defined quantity, but we have found that
its proper definition is not readily available, and worse, it is often incorrectly defined in polarization
work.

---

[39]This rotation does not occur for an equatorially mounted telescope. Unfortunately, the Hat Creek 85 ft and NRAO
140 ft telescopes were the last such large single-dish radio telescopes in operation; the former was destroyed during a
windstorm in 1993, and the latter, while removed from mothballs in 2005, is not being used for astrophysical research.

2.4.4.1   The Parallactic Angle

The parallactic angle of a source is defined as the angle between the north celestial pole (NCP) and the zenith measured from north toward east. The four-parts formula from spherical trigonometry (Thompson et al. 2001; van de Kamp 1967; Green 1985) can be used to show that the parallactic angle is defined for the geocentric equatorial and horizon (or alt-az) coordinate systems,[40] respectively, as:

$$\psi_{\mathrm{PA}} = -\tan^{-1}(-\sin(\mathrm{LST} - \alpha), \cos\delta\tan\mathcal{L} - \sin\delta\cos(\mathrm{LST} - \alpha)) \qquad (2.59\mathrm{a})$$

$$= -\tan^{-1}(\sin\mathcal{A}z, \cos\mathcal{E}l\tan\mathcal{L} - \sin\mathcal{E}l\cos\mathcal{A}z), \qquad (2.59\mathrm{b})$$

where LST is the local sidereal time, (α, δ) are the right ascension and declination of the source, $(\mathcal{A}z, \mathcal{E}l)$ are the azimuth and elevation of the source, and $\mathcal{L}$ is the latitude of the telescope.[41] When the source is in the east ($\mathrm{LST} < \alpha$), the parallactic angle will be negative; it will be positive when the source is in the west ($\mathrm{LST} > \alpha$). For sources north of the zenith, ψ_{PA} decreases with time and therefore rotates clockwise; for sources south of the zenith, ψ_{PA} rotates counterclockwise with time. For an alt-az telescope with linear feeds, the polarization of the feed is usually aligned with the direction to the zenith, i.e., along the meridian.

It can be shown easily from equations (2.59) that: as the zenith is approached, the parallactic angle reduces to $\psi_{\mathrm{PA}} = \mathcal{A}z - 180°$;[42] as the NCP is approached, $\psi_{\mathrm{PA}} = -(\mathrm{HA} - 180°)$, where the hour angle is $\mathrm{HA} = \mathrm{LST} - \alpha$; and as the south celestial pole is approached, $\psi_{\mathrm{PA}} = \mathrm{HA}$.

2.4.4.2   Beam Squint and Beam Squash

We mapped the beam response of the GBT using the bright continuum source Cas A. Figure 2.3 shows the Stokes I response of the telescope. The grayscale and contours show the fractional response relative to the peak Stokes I value. It is clear that the GBT has a symmetric main beam

---

[40]One should be careful to use equation (2.59a) because equation (2.59b) fails at the poles, which are singularities in the horizon coordinate system. It might be true that the zenith is a singularity in the equatorial coordinate system, but single-dish alt-az telescopes can't track sources directly through the zenith; however, observing the poles—especially the NCP, where the magnetic field is quite strong—is routine.

[41]The nomenclature $\psi = \tan^{-1}(y, x)$ is equivalent to the standard arctangent of y/x except that the quadrant in which ψ lies is computed, and the returned value will be in the range $-\pi < \psi \leq \pi$.

[42]One needs to be careful when using the Arecibo telescope since the value of azimuth stored in the Arecibo data structure is actually the *encoder* azimuth: the source's azimuth, because of the geometry of the stationary primary, will be 180° away from the stored azimuth. Heiles et al. (2001b) make this error on p. 1277 when they state that $\psi_{\mathrm{PA}} \sim \mathcal{A}z$ for a source near zenith: they are explicitly and incorrectly referring to the *source* azimuth. This error is repeated on p. 144 of Heiles (2002). It will be seen in § 2.4.4.2 that the linear response pattern will be insensitive to a 180° error in the parallactic angle, but the circular response pattern has a direction on the sky so that the correct definition of parallactic angle must be used.

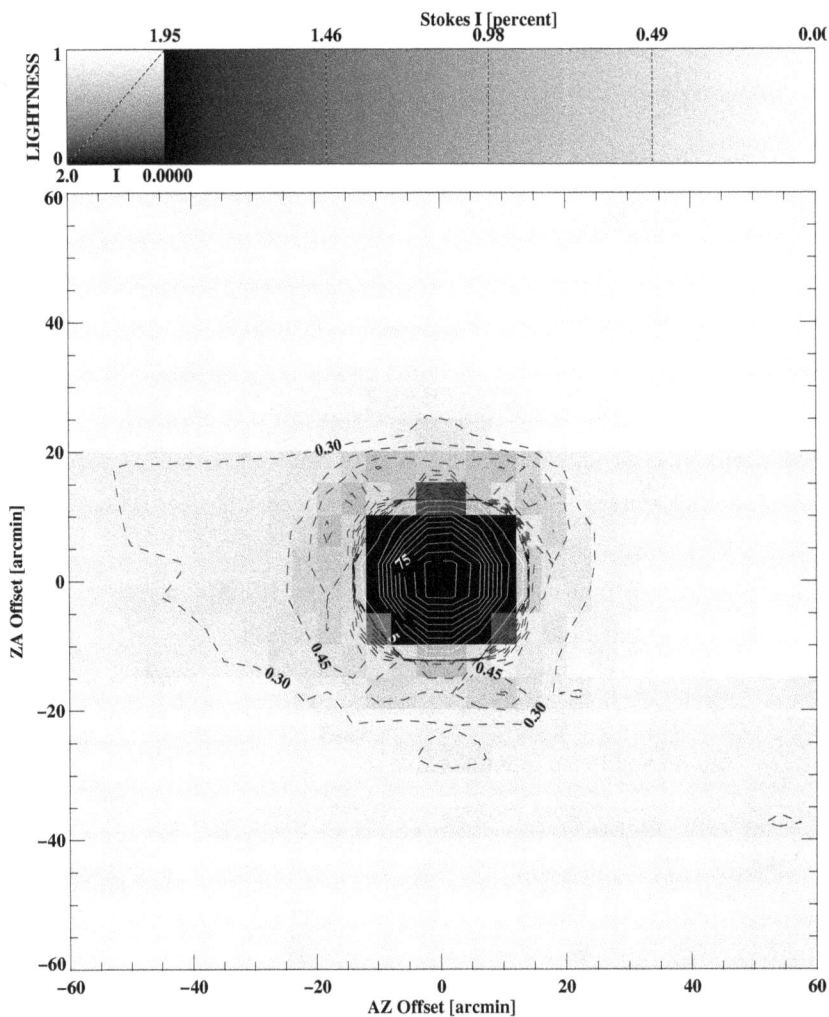

FIG. 2.3 — Total intensity Stokes I beam map for the GBT 100m. The grayscale and contours show the percentage of the peak response, with dashed black contours spaced by 0.15% and solid white contours spaced by 5%. The sidelobes of the GBT are symmetric, and the first sidelobe is suppressed ∼24 dB from the main beam response.

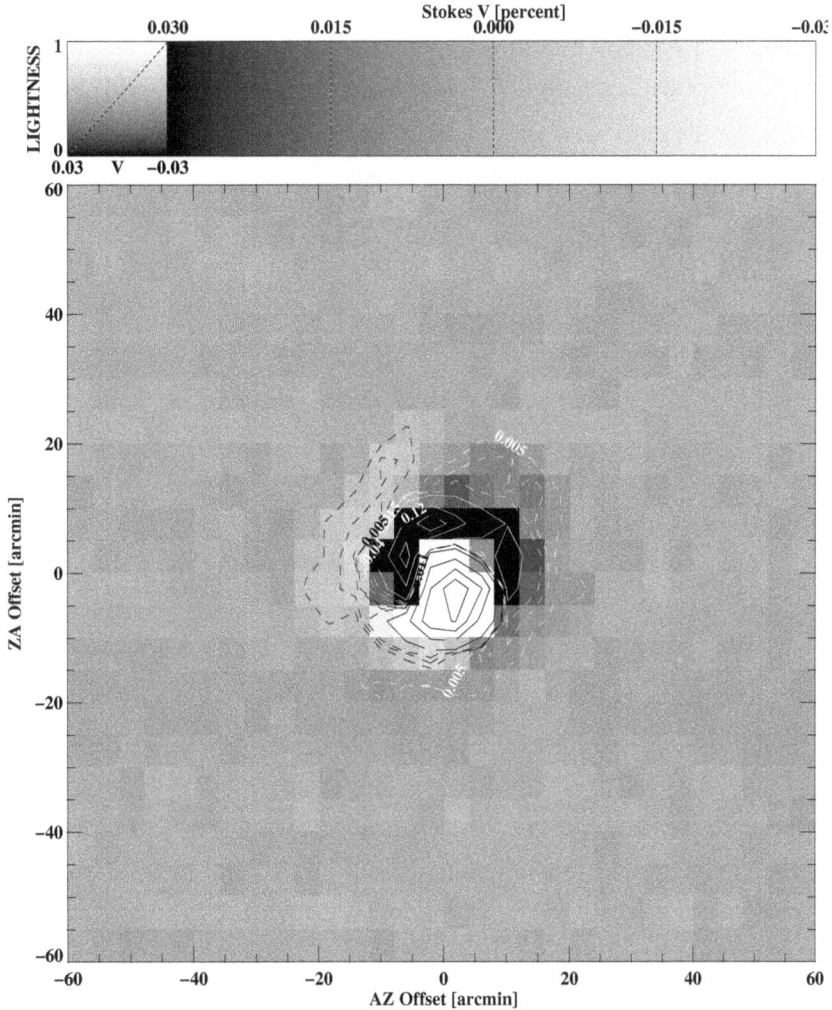

FIG. 2.4 — Circular polarization Stokes V beam map for the GBT 100m. The grayscale and contours show the percentage of the peak Stokes I response, with solid white and black contours showing the positive and negative response, respectively, inside the main beam. The dashed white and black contours show the positive and negative response, respectively, in the first sidelobe. The beam displays the expected squint pattern in both the main beam and the first sidelobe.

and that the first sidelobe is at a level of 0.5%. It was designed to minimize sidelobe response by placing the feed, and its supporting structure, off of the main axis of the telescope. It should be noted that the map is a convolution of the true telescope beam and the spatial distribution of Cas A, which is not a point source.

It is well known that an offset-parabolic telescope should have a circular polarization response characterized by *beam squint* (Rudge & Adatia 1978; Chu & Turrin 1973). Beam squint is caused by the LCP and RCP responses each being slightly offset from the symmetry axis (Rudge & Adatia 1978, Fig. 15). The Stokes V pattern is the difference between the RCP and LCP patterns. The result is a two-lobed pattern with the lobes symmetrically displaced on either side of the Stokes I main beam center. By mapping the Stokes V response of the GBT, we were able to quantify the beam squint; our result is shown in Figure 2.4. Amazingly, the expected sidelobe squint is seen as well; the Arecibo squint map (Heiles et al. 2001a) shows far too much structure in the sidelobes for this effect to be seen. The beam squint is typically quantified by the vector $\mathbf{\Psi}_B$ where the magnitude has been shown (Rudge & Adatia 1978) to be

$$\Psi_B = \arcsin\left(\frac{\lambda \sin\theta_0}{4\pi F}\right), \tag{2.60}$$

where F is the focal length of the primary, and θ_0 is the offset angle of the feed. The direction of the squint is measured from the mininum response in the V beam toward the maximum response. The maximum and minimum of the lobes will be separated by roughly the half-power beam width of the main beam, and the peak amplitude of the response will be proportional to Ψ_B. The peak-to-peak amplitude of the squint response in Figure 2.4 is $\sim$0.4%, which corresponds to a squint of $\Psi_B \approx 2''$.

For a source with a spatial extent much smaller than the main beam, the beam squint will not affect the measured circular polarization. However, if the Stokes I emission is extended compared to the main beam, then a fake Stokes V response can be produced. The squint pattern responds to the first derivative of the spectral line, i.e., the spatial gradient of Stokes I. Therefore, if there is a velocity gradient in the spectral line across the beam, and the squint response is aligned with this gradient, then the squint contribution to the Stokes V spectrum will be an S curve whose peaks are separated in velocity by $\nabla v \cdot \mathbf{\Psi}_B$ (Heiles 1996). If a source were tracked over a parallactic angle range of $\sim$180°, this instrumental contribution would average out of the final Stokes V spectrum.[43] However, since this lucky scenario rarely obtains, it is a far safer procedure to remove any parallactic angle dependence of the Stokes V spectra via least-squares fitting. Heiles & Troland (2004) and Heiles (1996) discuss the details of estimating and fitting for the instrumental contribution of the beam squint to the measured Stokes V response.

For completeness, we show the linear polarization beam response in Figure 2.5. Heiles et al. (2001a) coined the term *beam squash* to describe this four-lobed response. However, this pattern has been known to exist for offset parabolic reflectors for some time (see Napier 1999 for a nice

---

[43]The beam only rotates relative to the velocity structure for alt-az telescopes, which are the only remaining flavor of large single-dish telescope.

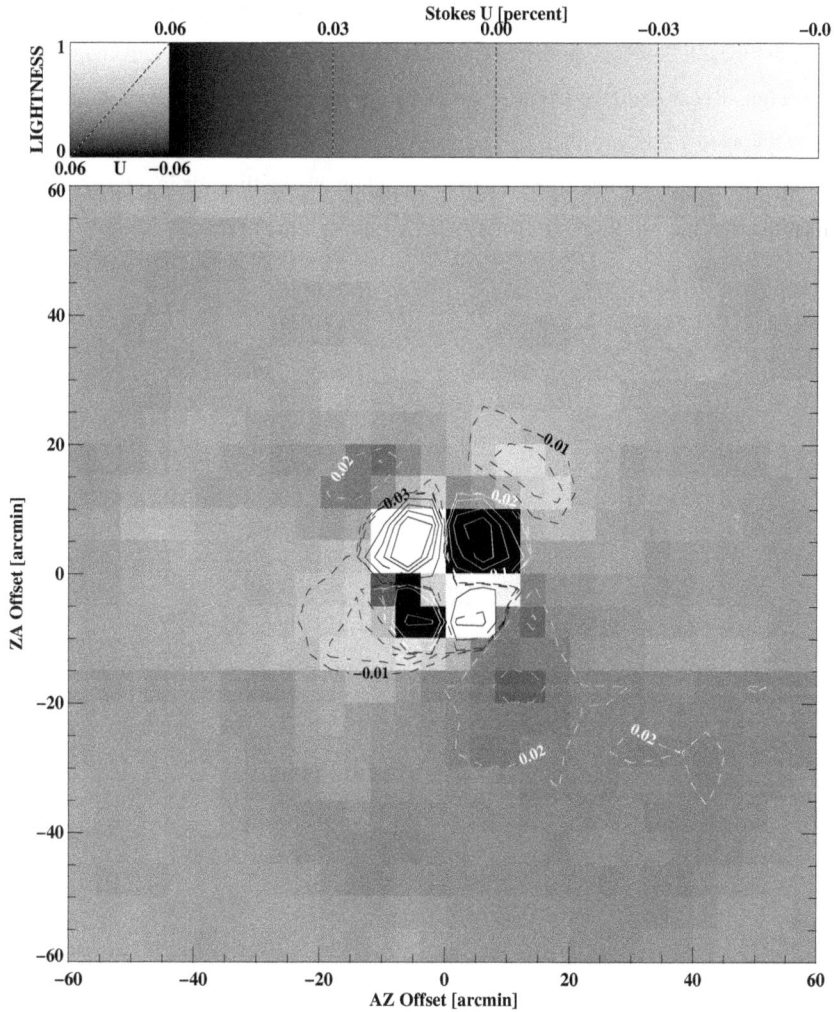

FIG. 2.5 — Linear polarization Stokes U beam map for the GBT 100m. The grayscale and contours show the percentage of the peak Stokes I response, with solid white and black contours showing the positive and negative response, respectively, inside the main beam. The dashed white and black contours show the positive and negative response, respectively, in the near sidelobes. The expected four-lobed squash response is seen in the main beam. Response is also seen in the first sidelobe and beyond.

discussion). There will be two lobes of the same sign on either side of the main beam center. The other pair of lobes will be orthogonal to the first set and will have a negative response. The squash can result from curvature effects of the primary reflector as well as from the non-circularity of the feed pattern. We point out that the sidelobe response is seen to change sign from that of the main beam squash lobes. It is not immediately clear whether this is to be expected from the standpoint of antenna design. Regardless, the linear polarization response will not affect our Stokes V spectra, so we do not consider it any further.

Heiles (1990) offers a final recommendation when accounting for instrumental polarization in Zeeman measurements: "In all of these matters it never hurts to pray."

Chapter 3

21 cm Zeeman Measurements in the Taurus Molecular Cloud: Constraining the Bull

> The outcome of any serious research can only be to make two
> questions grow where only one grew before.
>
> THORSTEIN VEBLEN, 1908

Abstract

Excess rotation measures have been detected by Wolleben & Reich (2004b) toward multiple polarized intensity minima in the Taurus Molecular Cloud (TMC) complex. By modeling each of these regions as a depolarizing magneto-ionic screen, they determine that the line-of-sight magnetic field in the TMC region is in excess of 20 μG. The proposed location of these Faraday screens is at the cloud surface, where 21 cm emission from the warm neutral medium at the boundary of the molecular cloud should probe the same $\gtrsim 20$ μG field and be easily observed via the Zeeman effect. We observed circularly polarized 21 cm emission toward three of these objects and report the clear detection of multiple magnetic fields. The scenario is complicated. We find strong fields of between $+35$ and $+40$ μG associated with a velocity component near -50 km s^{-1}. However, the TMC molecular emission is completely constrained to 0–10 km s^{-1}, implying that this ~ 40 μG field is not associated with the molecular cloud. On the other hand, we also detected fields with of strength $\sim +5$ μG toward each of the objects at a velocity of ~ 8 km s^{-1}, which is conincident with the velocity of the molecular emission. However, there is no unambiguous way to directly associate the H I emission at this velocity with the molecular component. Regardless, our observations do rule out the possibility of the existence of a line-of-sight field with strength $\gtrsim 20\ \mu$G in the TMC complex. We outline previous magnetic field estimates in the TMC region via multiple observational methods and discuss each in the context of our results. Finally, we suggest that the observed Faraday screens might be associated with the -50 km s^{-1} H I filament.

3.1 Introduction

Wolleben & Reich (2004a,b) have recently mapped a region of the Taurus Molecular Cloud (TMC) complex in the continuum at 1408, 1660, and 1713 MHz, and have discovered nine positions where the polarized intensities are minima and the rotation measures (RMs; calculated using the 1408 and 1660 MHz data) are excessive. Since there is no corresponding Stokes I emission, they interpret the Faraday rotation to occur in a magneto-ionic medium at the surface of the molecular clouds toward which they are seen; these Faraday screens will not radiate and will therefore have no total intensity fingerprint, but they will alter the linearly polarized background Galactic synchrotron emission.

Figure 2 of Wolleben & Reich (2004a) shows 3 adjacent regions of RM > 20 rad m^{-2} near Galactic coordinates $(\ell, b) = (169\overset{\circ}{.}5, 9\overset{\circ}{.}6)$. In a more detailed analysis over a wider area, Wolleben & Reich (2004b, henceforth WR04b) label these positions objects c–e, and their measured RMs have decreased to 1, 15, and 26 rad m^{-2}, respectively.[1] Faraday rotation allows an electron density-weighted estimate of the line-of-sight magnetic field in the screen, where the field can be inferred from the RM as:

$$\mathrm{RM} = 0.81 \left(\frac{B_\parallel}{\mu\mathrm{G}} \right) \left(\frac{n_e}{\mathrm{cm}^{-3}} \right) \left(\frac{l}{\mathrm{pc}} \right)\ , \tag{3.1}$$

where $B_\parallel$ is the line-of-sight magnetic field, n_e is the electron density in the magneto-ionic medium, and l is the length of the Faraday screen along the line of sight. This assumes that n_e is constant within the screen. The length l is estimated to be ~ 2 pc from the geometry of each screen under the assumption that it is located at the TMC distance of 140 pc. Using a simple model of the Faraday screens, and taking an upper limit of $n_e \lesssim 0.8$ cm^{-3}, they estimate a line-of-sight magnetic field of *at least* 20 μG is necessary at the cloud surface in order to explain the derived RMs.

This extraordinarily high field estimate has been cited in star formation studies (Bot et al. 2007; Sun et al. 2008) and mentioned as a new tool for probing the magnetic field structure at molecular cloud boundaries using the Square Kilometer Array (Beck & Gaensler 2004). However,

---

[1] Henceforth, we refer to these objects as WRc, WRd, and WRe.

TABLE 3.1

TARGET POSITIONS

Wolleben & Reich (2004b) Object	R. A. (J2000)	Dec (J2000)	ℓ (deg)	b (deg)
WRc.........................	04:42:44	31:09:36	170.23	−9.77
WRd	04:41:49	31:36:40	169.76	−9.62
WRe.........................	04:40:59	31:59:39	169.35	−9.51

other authors are dubious of the estimated field strength and have called for Zeeman measurements to confirm or constrain their interpretation (Nakamura & Li 2005; Price & Bate 2008).

We therefore aimed to directly probe the magnetic field at the surface of the molecular cloud by attempting to detect Zeeman splitting in the 21 cm radiation (either emission or self-absorption) toward WRc, WRd, and WRe, where a field of 20 μG would be easily detected in a few hours of integration. The observations are described in § 3.2. We discuss the data reduction in § 3.3 and the results in § 3.4. In § 3.5, we summarize the current knowledge of the magnetic field in the Taurus region as well as the methods by which these inferences were made. A summary is offered in § 3.6.

3.2 Observations

In 2005 February, we set out to detect the proposed $\gtrsim 20$ μG fields via Zeeman splitting of the 21 cm line using the 100 m Robert C. Byrd Green Bank Telescope[2] (GBT). Using the *L*-band receiver and the Spectral Processor in full-Stokes mode, we observed both the 1420 MHz H I and 1667 MHz OH transitions. Using a bandwidth of 2.5 MHz across 512 spectral channels, we employed in-band frequency switching to maximize efficiency. The integration time was 60 s and the switch period was 1 s. The three target positions are listed in Table 3.1. Rather than using the standard GBT calibration method involving a "winking" low-temperature ($\sim$ 1 K) noise source ("cal"), we observed for 10 min intervals with no cal, and then injected the high-temperature ($\sim$ 20 K) correlated cal for 10 s. The total integration times were 7 hr for WRc and 6.25 hr each for WRd and WRe. Polarization calibration was done via the now-standard GBT technique (Heiles et al. 2001b; Heiles & Troland 2004) of "spider scans" on the linearly polarized continuum source 3C 286. We calibrated the sense of circular polarization, and thus the direction of the magnetic field, by observing the 1665 MHz OH maser emission from both W49 and W3, and by comparing with the

---

[2]The National Radio Astronomy Observatory is a facility of the National Science Foundation operated under cooperative agreement by Associated Universities, Inc.

carefully calibrated results of Coles & Rumsey (1970), who employed the proper IEEE definition of right-handed circular polarization (RCP) and the IAU definition of Stokes $V \equiv \text{RCP} - \text{LCP}$.[3]

3.3   Data Reduction

The data were searched visually for radio frequency interference (RFI); all spectra containing insidious RFI were completely rejected, while narrow RFI spikes were interpolated over. The gain was determined in 10 min intervals by injection of the correlated cal; the derived gain was applied to the following cal-off spectra. The phase and gain differences between the two polarization channels were corrected for via calibration using the correlated cal (see Heiles et al. 2001b; Heiles 2001).

By mapping a linearly polarized continuum source, we have found that the Stokes V response of the GBT displays the expected beam squint structure (see Heiles 1996; Heiles & Troland 2004). We follow Heiles (1996) in using an empirical estimation of the contribution to the Stokes V spectrum from the beam squint. For a given spectral channel, we performed a least-squares fit to our calibrated and Mueller matrix-corrected Stokes V data of the form

$$V_i(t) = A_i + B_i \sin \psi_{\text{PA}}(t) + C_i \cos \psi_{\text{PA}}(t), \qquad (3.2)$$

where $V_i(t)$ represents the ith channel of every Stokes V spectrum, $\psi_{\text{PA}}(t)$ is the parallactic angle of the source at the time each spectrum was obtained, the fitted parameters B_i and C_i represent the squint contribution to Stokes V, and the constant term A_i is the true, instrumentally-independent Stokes V contribution for the spectral channel i. Our final Stokes V spectrum is the composite A_i spectrum.

3.4   Results

Figures 3.1–3.3 show the averaged Stokes I and V spectra for each of our target positions. The top panel of each plot shows the Stokes I spectrum as a solid line, the components of a Gaussian decomposition to the Stokes I profile as dashed lines, the expanded residuals of the profile from the composite Gaussian fit as a thin solid line at the center height of the plot, and a scale bar for the expanded residuals to the right of the plot. The zero-based number of each Gaussian component

---

[3]The IAU convention for Stokes V had not yet been established, but their choice would later prove to fall in line with the IAU (1974) convention. A search of the literature on circular polarimetric radio observations shows that the IAU convention has largely been unknown or ignored for the last 35 years: it is rarely cited, and V is quite often presented as LCP − RCP.

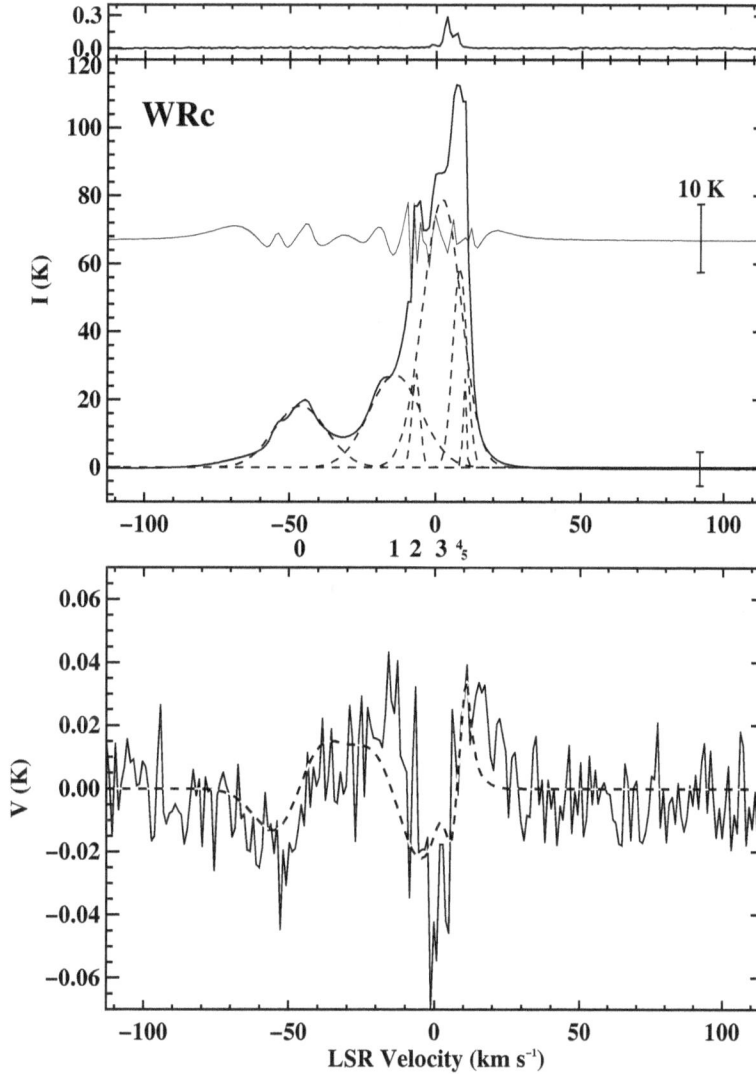

FIG. 3.1—Stokes I and Stokes V results for WRc. *Top*: 21 cm Stokes I spectrum (*solid line*; the sum of the two orthogonal feed responses, not the conventionally plotted average) as a function of LSR velocity. The profile of each Gaussian component is plotted as a dashed line, with the corresponding zero-based component number shown below the velocity axis at the corresponding central velocity. Residuals from the composite Gaussian fit are plotted through the center of the panel (*thin solid line*) and are expanded by a factor of 2. The scale bars near the right edge of the plot correspond to the labeled temperature range. The spectrum plotted above the 21 cm profile is the 1667 MHz OH Stokes I profile. *Bottom*: 21 cm Stokes V spectrum (*solid line*; RCP − LCP) and its fit (*dashed line*).

TABLE 3.2

WRC GAUSSIAN FIT PARAMETERS

Gaussian (1)	I (K) (2)	v_{LSR} (km s^{-1}) (3)	Δv (km s^{-1}) (4)	$B_{\parallel}$ (μG) (5)
0	18.17 ± 0.38	-46.35 ± 0.29	21.23 ± 0.73	36.45 ± 9.24
1	27.17 ± 1.29	-13.83 ± 1.15	21.05 ± 1.89	-20.51 ± 6.60
2	28.06 ± 1.23	-6.64 ± 0.06	3.15 ± 0.17	0.09 ± 2.59
3	79.01 ± 2.83	2.13 ± 0.32	15.44 ± 0.36	5.23 ± 1.91
4	0.59 ± 1.88	8.25 ± 0.07	5.59 ± 0.17	5.19 ± 1.52
5	26.96 ± 1.83	10.00 ± 0.07	1.26 ± 0.13	1.10 ± 2.80

TABLE 3.3

WRD GAUSSIAN FIT PARAMETERS

Gaussian (1)	I (K) (2)	v_{LSR} (km s^{-1}) (3)	Δv (km s^{-1}) (4)	$B_{\parallel}$ (μG) (5)
0	14.62 ± 0.25	-45.57 ± 0.38	26.05 ± 0.86	45.83 ± 11.85
1	4.48 ± 0.68	-18.83 ± 0.33	5.02 ± 0.95	8.70 ± 17.10
2	28.68 ± 0.97	-6.34 ± 0.56	29.57 ± 0.69	31.67 ± 6.82
3	9.31 ± 1.05	-5.63 ± 0.17	4.21 ± 0.49	-17.72 ± 8.14
4	0.78 ± 1.35	2.66 ± 0.10	11.75 ± 0.30	2.90 ± 1.54
5	65.24 ± 1.64	8.53 ± 0.02	4.25 ± 0.09	7.56 ± 1.10

is placed just below the bottom axis at its corresponding velocity; the fitted parameters for each Gaussian are listed in Tables 3.2–3.4. The 1667 MHz OH Stokes I profile is shown in a panel directly above the 21 cm profile. The bottom panel of each plot shows the Stokes V spectrum as a solid line and the best-fit Zeeman profile as a dashed line.

It is clear from each of the Stokes V spectra in Figures 3.1–3.3 that there is a detectable circular polarization toward each position. The spectral fingerprint of Zeeman splitting in the Stokes V spectrum is a feature that is proportional to the derivative of the Stokes I profile (this is often referred to as the S curve) and one can least-squares fit the Stokes V spectrum with the functional form:

$$V = \frac{1}{2}\frac{dI}{d\nu} b B_{\parallel},$$
(3.3)

where b is the Zeeman splitting coefficient for the 21 cm line, for which $b = 2.8$ Hz μG^{-1}, $dI/d\nu$ is the derivative of the classically defined Stokes I spectrum (i.e., not the average of the two linear feed responses, but the sum), and $B_{\parallel}$ is the line-of-sight magnetic field in μG. Since the circular polarization features cover the entire extent of the Stokes I emission, which is comprised of many narrow warm-neutral medium (WNM) components, in addition to extremely narrow cold-neutral

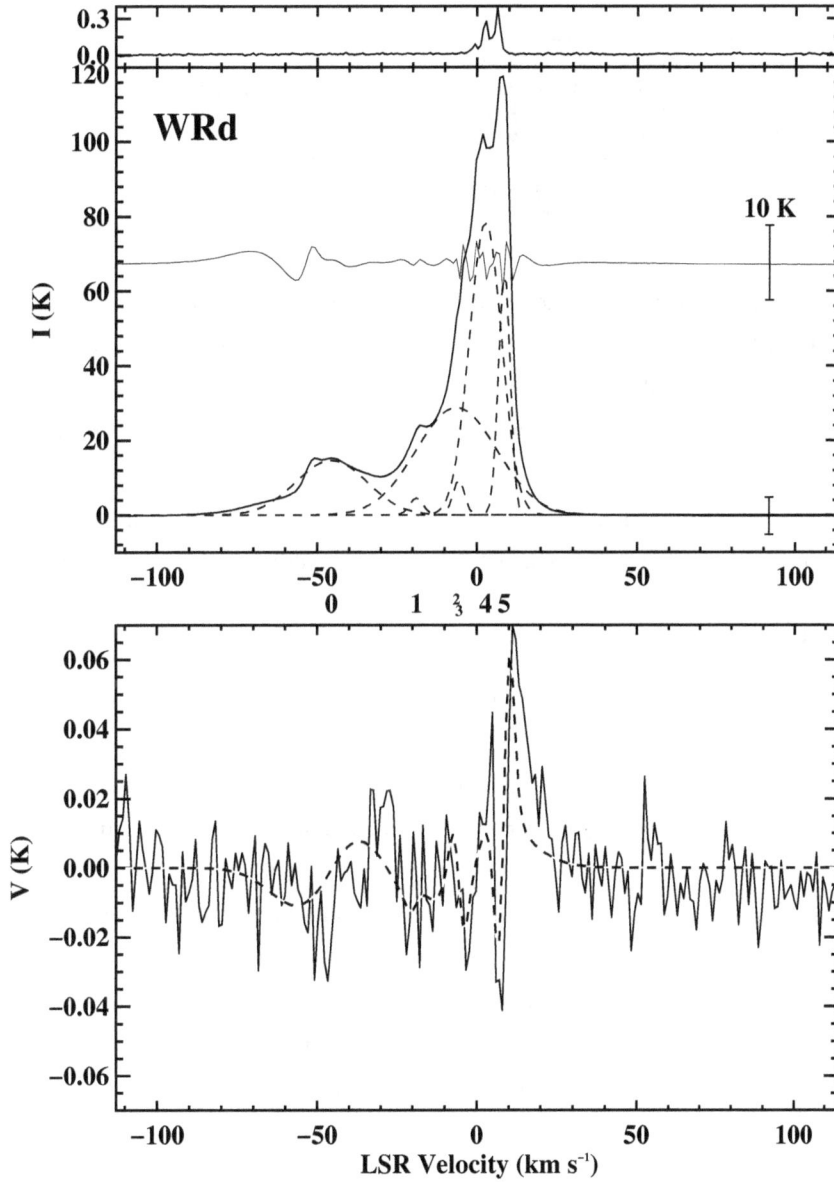

FIG. 3.2 — Stokes I and Stokes V results for WRd. See caption for Fig. 3.1.

TABLE 3.4

WRE GAUSSIAN FIT PARAMETERS

Gaussian (1)	I (K) (2)	v_{LSR} (km s^{-1}) (3)	Δv (km s^{-1}) (4)	$B_{\parallel}$ (μG) (5)
0	23.30 ± 0.36	-44.93 ± 0.20	19.29 ± 0.47	39.77 ± 7.76
1	3.07 ± 0.84	-18.67 ± 0.63	5.05 ± 1.79	13.00 ± 30.94
2	6.16 ± 1.05	-8.16 ± 0.21	2.69 ± 0.58	-25.21 ± 13.03
3	24.40 ± 1.69	-7.15 ± 0.76	31.43 ± 0.92	23.46 ± 10.12
4	0.72 ± 1.86	1.78 ± 0.11	14.04 ± 0.34	12.97 ± 2.24
5	61.84 ± 1.31	8.44 ± 0.03	3.98 ± 0.09	5.50 ± 1.41

medium (CNM) components, all at various velocities, and therefore distances, along the line of sight, we cannot fit for a single field. The Zeeman effect probes the in situ field at the location of the emission, so we need to fit each emission component with a separate S curve in order to estimate multiple field strengths. Without physical motivation for choosing components, the best that we can do is to fit the Stokes I spectrum with as few components as possible that yield reasonable residuals (see the discussion of § 4.4.4). The parameters and formal errors of these fits and the derived line-of-sight magnetic fields (strengths and directions) are presented in Tables 3.2–3.4.

It is encouraging to note that, for a given object, the Stokes V spectral shapes appear identical when analyzing data taken one week apart; also, performing our full reduction on either the signal or reference frequency-switch phase alone provides the same Stokes V shape.

Our 1667 MHz OH spectra show emission ranging from -2 to $+10$ km s^{-1}. This is co-incident with the velocity extent of the high-resolution ^{12}CO and ^{13}CO emission (0–12 km s^{-1}) integrated over many square degrees in the TMC region (Goldsmith et al. 2008). While there are OH peaks near the velocities of local minima in the H I profiles toward WRd and WRe, Gaussian fits that included an opacity component to model narrow self-absorption features failed to converge. Therefore we are unable to associate any Zeeman signature with a cold neutral medium (CNM) absorption component, which would have allowed us to probe the field in the molecular region. Our total integration time was far too small to attain any sensitivity for Zeeman splitting of the OH line, so we do not show the Stokes V spectra. For completeness, a single-field fit to the Stokes V spectra (with $b = 1.96$ Hz μG^{-1}) produced formal 1 σ errors of $\sim$80 μG.

We are left only with the fields associated with the multiple 21 cm emission features that are spread out significantly in velocity. It is tempting to suggest that the components near 8.5 km s^{-1} in each spectrum are all associated with the molecular emission, perhaps WNM components

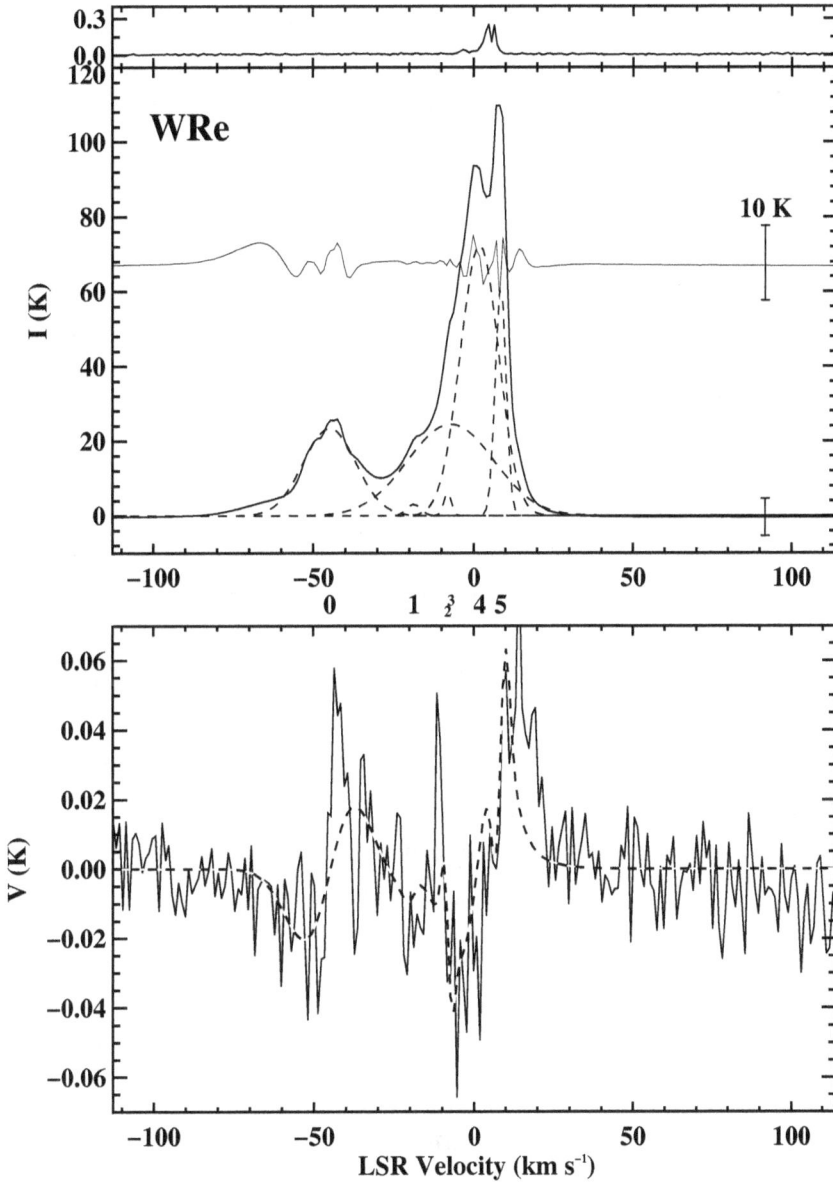

FIG. 3.3 — Stokes I and Stokes V results for WRe. See caption for Fig. 3.1.

(the FWHM of each component ranges from 4 to 6 km s^{-1}) at the periphery of the molecular cloud. We would therefore be probing a line-of-sight field of roughly +5 μG at the surface of the molecular cloud. To investigate this possibility, we used the all-sky 21 cm LAB survey (Kalberla et al. 2005; Hartmann & Burton 1997) to make an image of the H I column density integrated over the range 5–9 km s^{-1}. This image is shown in Figure 3.4 with CO contours from Dame et al. (2001). Out target positions are shown as diamonds in the northwest. There is no obvious correlation between the H I and the CO structure over these channels in the TMC region. One can make the general observation from Figure 3.4 that there is a gradient in the H I column density from northeast to southwest, which also appears to be the orientation of one of the principal axes of the CO distribution. We conclude that there is no way to directly associate our estimated field strengths with the molecular emission in the TMC region. Further, we note that it would be useful to map the circular polarization of this region in order to discover if the WNM component near 8 km s^{-1}, which is present across this complex, traces a diffuse magnetic field of microgauss strength.

As for the $\sim$+40 μG field detections in the components near -50 km s^{-1}, Figure 3.5 shows the LAB H I column density integrated from -55 to -40 km s^{-1}. It can be seen that our targets lie on the edge of an extended filamentary structure with a length of almost 15°. This filament is adjacent to another that is extremely straight and extends from $(\ell, b) = (166°, -9°)$ to $(\ell, b) = (182°, -17°)$; its length-to-width ratio is $\approx$12. It is not inconceivable that the shape of these filaments is constrained by magnetic fields. It would seem that observing a single slice in latitude across both filaments would allow us to probe the field strength at this velocity in order to see if it is enhanced inside the filaments. If so, a map of the Zeeman splitting in combination with plane-of-sky magnetic field estimates from starlight polarization might allow for a three-dimensional field to be modeled in this region. If one plots the positions of the remaining Wolleben & Reich (2004b) Faraday screens, it is abundantly clear that there is a direct overlap with this filamentary structure; the correlation of these positions with the -50 km s^{-1} H I emission is as strong, or stronger, than their correlation with the TMC molecular emission.

3.5   Magnetic Field Estimates in the TMC Complex

Over the last decade there have been a number of estimates of the magnetic field strength in the TMC region. Below we shall outline the details of each and attempt to contrast each method with our Zeeman-splitting results.

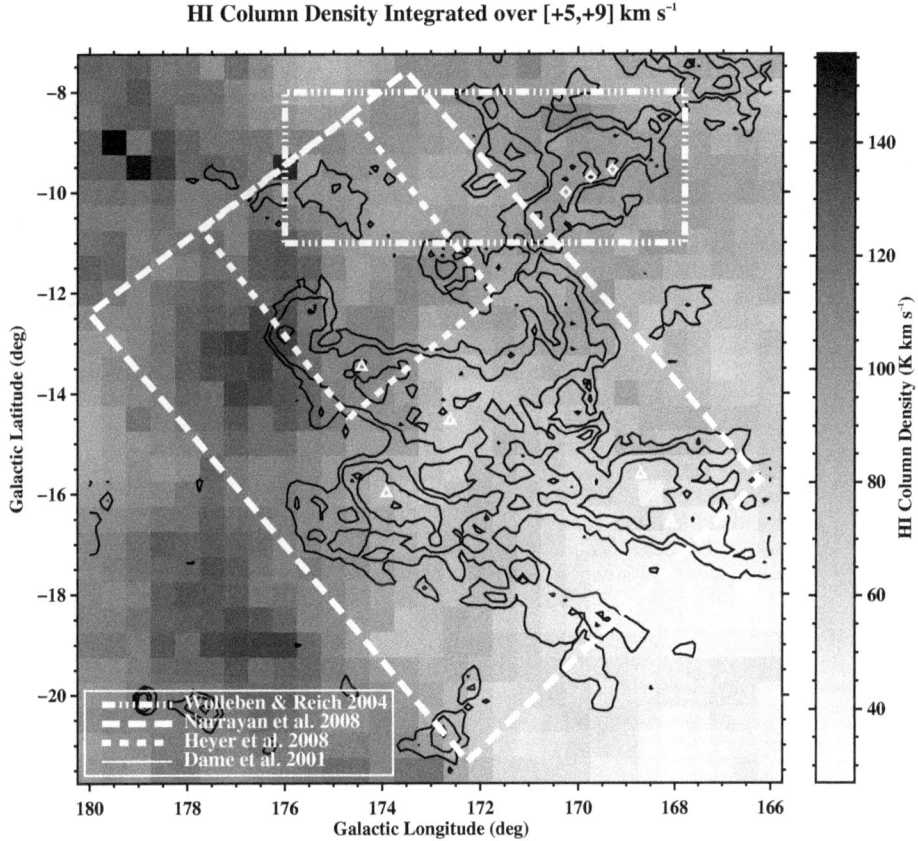

HI Column Density Integrated over [+5,+9] km s^{-1}

FIG. 3.4 — Neutral H I and molecular CO content in the vicinity of the TMC complex. The grayscale image shows the LAB H I emission (Kalberla et al. 2005) intergrated from 5 to 9 km s^{-1}. The contours represent the Dame et al. (2001) integrated CO data. Three boxes are outlined: the region mapped in CO by Narayanan et al. (2008) (*long dashes*) using the FCRAO 14 m, the region analyzed by Heyer et al. (2008) (*short dashes*) to estimate field strengths of ~14 μG using the FCRAO data set, and the region mapped by Wolleben & Reich (2004b) using the Effelsberg 100 m (*dot dashes*). The three diamonds represent our target positions, locations where Wolleben & Reich (2004b) estimate field strengths $\gtrsim 20$ μG. The five triangles are positions where Crutcher et al. (1993) found limits of $|B_\parallel| \lesssim 10$ μG.

HI Column Density Integrated over [−55,−40] km s⁻¹

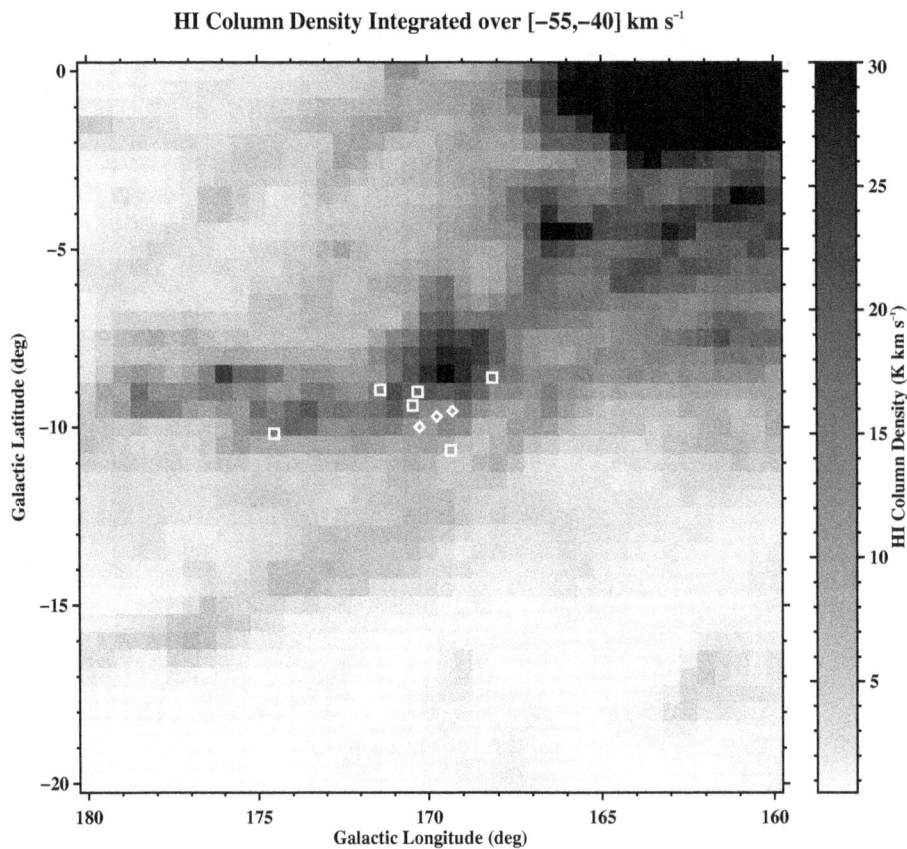

FIG. 3.5 — H I column density integrated from −55 to −40 km s⁻¹ in the TMC region. Two long, narrow filaments are seen running parallel to one another from the northwest to the southeast. The positions of our three targets are plotted as white diamonds, and those of the remaining WR04b Faraday screens are plotted as white squares. Each Faraday screen is located along a line of sight that intercepts these filaments.

3.5.1   The Zeeman Effect

The most reliable method for investigating the magnetic field is the Zeeman effect because it directly probes the field at the source of emission or absorption. There have only been a few previous Zeeman detections in the TMC region. Güsten & Fiebig (1990) initially reported a field of $B_\parallel = -100\,\mu$G using the 11.112 GHz CCS transition toward TMC-1C, but subsequently attributed this detection as spurious and the result of instrumental error (Troland et al. 1996). Crutcher et al. (1993) observed 5 positions in the TMC finding only upper limits of $|B_\parallel| \lesssim 10\,\mu$G. Using the 1665 and 1667 MHz OH emission lines, Troland et al. (1996) report a field limit of $B_\parallel = 1.4 \pm 2.4\,\mu$G toward TMC-1C. Crutcher & Troland (2000) observed 3 more cores in the TMC finding limits of $\lesssim 7\,\mu$G in two, and a clear detection of $B_\parallel = +11 \pm 2\,\mu$G in L1544, which is located at $(\ell, b) = (179°20, -6°24)$, just to the north of the northeastern corner of Figure 3.4. The positions sampled by the above Zeeman surveys are spread throughout the cores of the TMC and represent density enhancements where one would expect to see the line-of-sight field amplified relative to the ambient field in the TMC. Therefore, a detection of only $+11\,\mu$G in a dense core, as well as limits of $\sim 10\,\mu$G in a number of other dense cores, suggests that a $+20\,\mu$G field outside of the cores is very unlikely. In comparison to our Zeeman detections near $+8$ km s^{-1}, note that not only are the field strengths comparable, we also detect the same field *direction* pointing away from the observer.

3.5.2   Faraday Screens

We have outlined the region mapped by WR04b as a dash-dotted line in the northwestern region of Figure 3.4. It is clear that the positions of our targets are aligned with the edge of the molecular CO emission in this region. However, it should be noted that three of the seven Faraday screens that they have discovered (objects f, h, and i) are not convincingly aligned with CO concentrations: there is no certainty that the screens are associated with the molecular emission. Likewise, we are not able to state with certainty that any of our H I emission components is spatially adjacent to the molecular cloud. We therefore cannot be sure that the regions for which WR04b find Faraday screens, and for which we find Zeeman splitting of the 21 cm emission, are spatially coincident along the line of sight. However, as WR04b point out, we expect to see WNM emission in the outer regions of the molecular cloud: if the Faraday screens are indeed sampling fields of strength $\geq 20\,\mu$G, we would have detected it. If we were to associate our WNM component at 8 km s^{-1} with the outskirts of the molecular cloud, then we could claim a field strength of $+5$

μG, and then, assuming that their estimate for the 2 pc depth of the Faraday screen is sanguine, we could constrain the electron density of their model to be in line with 3.5 cm^{-3}.

Recently, Crutcher & Troland (R. M. Crutcher 2008, private communication) have used Arecibo to make Zeeman observations of OH emission toward WRd; they made no field detection and found a formal result for the line-of-sight field strength of 5$\pm$5 μG. This result, for which the OH emission is clearly associated with the molecular cloud, precludes the magnetic field strengths of $\gtrsim$ 20 μG proposed by WR04b. However, it does not contradict our possible detection of a +5 μG field in the TMC region.

WR04b measure positive RM toward each of the three objects that we have observed. The path length l in equation (3.1) is defined to be positive in the direction of the observer, but the convention for the sign of a magnetic field is that a positive field points away from the observer. Therefore, positive RM implies a negative field, or a field pointing toward the observer. While WR04b never infer a field direction from their positive RM observations, it appears that the Faraday screens are probing fields that are pointing toward us. Our detections indicate fields that are pointing away.

In obtaining the 20 μG lower limit, Wolleben & Reich (2004b) use an upper limit of $n_e = 0.8$ cm^{-3} obtained from the absence of excess Hα emission. This upper limit assumes an electron temperature of $T_e = 8000$ K. However, the electrons are most likely located in the predominantly neutral gas rather than in the ionized gas. The H I temperature may easily be as low as $\sim$100 K. The H II recombination coefficient increases with decreasing temperature, roughly as a power law with index -0.86 (Spitzer 1978, Table 4.5). Retaining their assumed path length of $l \approx 2$ pc, the excess electron density, including the temperature dependence, is $\Delta n_e \leq 0.8 T_8^{0.43}$, where $T_8 = T/8000$ K. If $T = 100$ K, then the upper limit on n_e becomes only 0.12 cm^{-3}, leading to $B \geq 3.75$ μG. Thus, including the temperature dependence drastically changes the expected field strength to a lower estimate that is in line with our Zeeman-splitting-derived result of $B \approx 5$ μG. Our Zeeman measurement is in line with the typical CNM field strength of 6 μG obtained by Heiles & Troland (2005). This is comforting, because the overall relationship between field strength and volume density shows that the field is independent of density up to a threshold $n_{HI} \sim 10^4$ cm^{-3} (R. M. Crutcher 2008, private communication). Volume densities in the non-core regions of the TMC are unlikely to exceed this threshold by very much, so this relationship agrees with our measured field strength being reasonably consistent with that in the typical CNM.

3.5.3 Velocity Anisotropy in the TMC

Measuring velocity anisotropy in spectral line data cubes is a relatively new method for estimating magnetic field strengths in the interstellar medium. Heyer et al. (2008) employ principal component analysis to detect anisotropy in both MHD simulations and actual astronomical data. Using a subfield (outlined in Fig. 3.4 using short dashes; see also Fig. 21 of Goldsmith et al. 2008) of the large-scale, high-resolution ^{12}CO $J = 1$–0 map (Narayanan et al. 2008, the entire extent of the survey is outlined by long dashes in Fig. 3.4) of the TMC[4] to estimate a minimum total magnetic field strength of 14 μG.

3.5.4 The Chandrasekhar-Fermi Method

A very popular way to estimate the plane-of-sky field strength is to employ the Chandrsekhar-Fermi method (Chandrasekhar & Fermi 1953, henceforth CF method). If one can measure the polarization distribution on the sky, either by means of starlight or dust emission, in addition to the velocity dispersion along the line of sight via a spectral line, then one can estimate the plane-of-sky magnetic field strength using (Zweibel 1990):

$$B_\perp^2 = f^2(4\pi\rho)\sigma_v^2/\sigma_{\rm pol}^2 \, , \tag{3.4}$$

where $\sigma_{\rm pol}$ is the dispersion of the measured polarization angles in radians, f is a factor accounting for density inhomogeneities, σ_v is the line-of-sight velocity dispersion, ρ is the mean density of the gas, and $B_\perp$ is the plane-of-sky projection of the mean magnetic field. A number of numerical simulations (Ostriker et al. 2001; Heitsch et al. 2001; Padoan et al. 2001) have been used to estimate that f is approximately one half.

Myers & Goodman (1991) compiled the dispersion angles from stellar polarization in the TMC complex and found that, while the dispersion was small, the mean magnetic field direction changed over the cloud. Crutcher et al. (2004) used the CF method on 850 μm polarized emission from SCUBA toward the dense core L1544, and found the plane-of-sky field to be 140 μG. Heyer et al. (2008) use the CF method on stellar polarization orientations within the same subfield (outlined in Fig. 3.4 by short dashes) of the TMC within which they employed the velocity anisotropy method. The high-resolution FCRAO map shows long hairlike molecular filaments that are remarkably well aligned with the plane-of-sky magnetic field orientation as revealed by optical polarization of starlight. By applying the CF method, Heyer et al. (2008) estimate a plane-of-sky

---

[4]The mapped region of WR04b intercepts the FCRAO field (Narayanan et al. 2008; Goldsmith et al. 2008) in the northwestern corner, but only one of their seven screens falls in this region.

field $B_\perp = 14 \ \mu$G, which is consistent with the results of their anisotropy study. This region is spatially adjacent to, and only a few degrees away from, our three targets. It should be noted that Heyer et al. (2008) use a value of $f = 0.5$, but as Crutcher et al. (2004) suggest, in a "diffuse region comprising dense filaments and cores as well as dilute gas," one ought to be using $f \approx 0.3$–0.4 according to findings from Heitsch et al. (2001). This would slightly lower the plane-of-sky field estimate in the northern region of Figure 3.4 to $\approx$8–11 μG.

3.6   Summary

We have observed Zeeman splitting in multiple H I emission features toward three lines of sight that are coincident with both a molecular cloud in the TMC complex and depolarizing Faraday screens discovered by WR04b. WR04b suggest that these screens are located at the periphery of this molecular cloud, and that a line-of-sight magnetic field of $\gtrsim$20 μG threads the magneto-ionic medium in these screens. Zeeman splitting of 21 cm radiation at a velocity coincident with that of the molecular emission from the TMC suggests that, if the neutral gas is associated with the envelope of the molecular cloud, a field of $\sim$+5 μG is present. These results preclude a line-of-sight field strength of 20 μG, and, if the Faraday screens are truly at the TMC distance, allow us to constrain the electron density of the modeled screens to a value $\sim$3.5 cm^{-3}.

The more remarkable result from our study is the $\sim$+40 μG field associated with the $\sim$−50 km s^{-1} H I filament. The magnetic pressure in this filament is $\sim$4 $\times$ 10^5 cm^{-3} K; how can such a pressure be maintained? The position of this object is near the Galactic anticenter; Galactic rotation is negligible in this vicinity, so we must attribute the velocity of −50 km s^{-1} to dynamical motions. The ram pressure of gas moving at 50 km s^{-1} is $\rho v^2/k \sim 4 \times 10^5$ cm^{-3} K. The equality of these two pressures suggests strongly that the filament and its field are transitory features associated with shocks. Two extreme shock models are those of the adiabatic and isothermal shock. The adiabatic model cannot apply here because a 50 km s^{-1} shock would ionize the gas; the presence of H I eliminates this possibility. The isothermal model makes much more sense. In this case, the field strength increases by the ratio of the post- to pre-shock density. The typical field strength in the ISM is perhaps 5 μ G. The field strength ratio should therefore be about $40/5 = 8$, which is also the density ratio. In a limiting case where the Alfvén velocity in the pre-shock medium much exceeds the sound velocity, we have (Spitzer 1978, eq. [10-27]):

$$\frac{\rho_2}{\rho_1} = \sqrt{2} \, \frac{u_1}{V_{\mathrm{A1}}}, \tag{3.5}$$

where the subscripts 2 and 1 are post- and pre-shock, respectively. For a shock velocity of 50 km s^{-1} and $\rho_2/\rho_1 = 8$, the Alfvén velocity $V_{A1} = 9$ km s^{-1}. For $B_1 = 5$ μG, the pre-shock H I density is $n_1 = 1.2$ cm^{-3} ,and the post-shock density is $n_2 = 9.6$ cm^{-3}. The column density of the -50 km s^{-1} component is $N(\text{H I}) \sim 4 \times 10^{20}$ cm^{-2}, so the line-of-sight thickness is $l = N(\text{H I})/n_2 \sim 7$ pc. This filamentary-like structure is about 1° wide; if its width is equal to l, then the distance to the filament is $\sim$400 pc. At a Galactic latitude of $-10°$, this makes the height $z \sim 70$ pc, well within the H I layer. This set of descriptive numbers is not at all unreasonable from a physical standpoint. From an astronomical standpoint, we might ask what could produce such a shock, and the usual answer is a supernova; we see of no reason to look further.

Suppose we have a 50 km s^{-1} shock and a pre-shock density of 1.2 cm^{-3}. Then immediately-shocked gas will be at least partially ionized and will produce a face-on Hα surface brightness of $I_{\text{H}\alpha} \approx 2.4$ R (Tufte et al. 1998, p. 783). The observed upper limit on Hα emission in this region is only 0.52 R. This is consistent with our above discussion: the gas has had time to cool forcing the shock to approach being isothermal. It is also consistent with the possible existence of pockets of partially ionized gas that would cause Faraday rotation. With our observed 40 μG post-shock field, the $B_{\parallel} \geq 20$ μG magnetic field inferred from the Faraday rotation data of WR04b is in complete agreement.

We cannot be sure, because the distance to neither the filament nor the Farday screens is known, but the assembled observational evidence suggests that the excess RMs found by WR04b might be associaated with the ~-50 km s^{-1} H I filament, rather than a molecular cloud in the TMC region.

These detections should be confirmed using Arecibo to ascertain whether our estimates are independent of instrumental effects, and if repeatable, Zeeman mapping should be performed in this region to quantify the spatial distribution of the line-of-sight field in both the TMC and the -50 km s^{-1} H I filament.

Acknowledgments

We would like to thank Toney Minter and Rick Fisher for help in preparing our observations. It is also a pleasure to thank Eric Knapp, Greg Monk, Kevin Gum, Dave Rose, Donna Stricklin, and Barry Sharp for their assistance and company while observing. This research was supported in part by NSF grant AST-0406987. Support for this work was also provided by the NSF to TR through awards GSSP 05-0001, 05-0003, and 05-0004 from the NRAO. This research has made use of NASA's Astrophysics Data System Abstract Service.

Chapter 4

Extragalactic Zeeman Detections in OH Megamasers

> I find that a great part of the information I have was acquired by looking up something and finding something else on the way.
>
> FRANKLIN P. ADAMS, 1960

A version of this chapter will be published in *The Astrophysical Journal* (Robishaw, Quataert, & Heiles 2008, ApJ, Vol. 680, Num. 2). ©2008 by the American Astronomical Society. All rights reserved. Reprinted by permission.

Abstract

We have measured the Zeeman splitting of OH megamaser emission at 1667 MHz from five (ultra)luminous infrared galaxies ([U]LIRGs) using the 305 m Arecibo telescope and the 100 m Green Bank Telescope. Five of eight targeted galaxies show significant Zeeman splitting detections, with 14 individual masing components detected and line-of-sight magnetic field strengths ranging from $\simeq 0.5$ to 18 mG. The detected field strengths are similar to those measured in Galactic OH masers, suggesting that the *local* process of massive star formation occurs under similar conditions in (U)LIRGs and the Galaxy, in spite of the vastly different large-scale environments. Our measured field strengths are also similar to magnetic field strengths in (U)LIRGs inferred from synchrotron observations, implying that milligauss magnetic fields likely pervade most phases of the interstellar medium in (U)LIRGs. These results provide a promising new tool for probing the astrophysics of distant galaxies.

4.1 Introduction

(Ultra)luminous infrared galaxies ([U]LIRGs) are a population of galaxies that emit far-infrared (FIR) radiation with energies comparable to those of the most luminous quasars [$\log(L_{\mathrm{FIR}}/L_\odot) > 11$ and 12 for LIRGs and ULIRGs, respectively]. Nearly every ULIRG appears to have undergone a merger/interaction and contains massive star formation and/or an active galactic nucleus (AGN) induced by gravitational interactions. Lo (2005) details very long baseline interferometry (VLBI) observations of the 1667 MHz hydroxyl (OH) transition in the nuclear regions in (U)LIRGs that have revealed multiple masing regions with $1 < \log(L_{\mathrm{OH}}/L_\odot) < 4$; these regions are known as OH megamasers (OHMs). Each OHM has a spectral line width of between 50 and 150 km s^{-1}; when viewed by a single dish, these spectral components are superimposed. The 1667 MHz OHM flux density is always a few to many times that of the 1665 MHz transition, and in many cases the 1665 MHz line is absent (Darling & Giovanelli 2002); this is an interesting contrast to the case of OH masers in the Galaxy in which the 1665 MHz transition is usually dominant (Reid & Moran 1988). The starbursts and AGNs in ULIRGs create strong FIR dust emission, as well as a strong radio continuum; the OHMs are generally believed to be pumped by the FIR radiation field (e.g., Randell et al. 1995) although collisional excitation may be important as well (e.g., Lonsdale et al. 1998). Lockett & Elitzur (2008) have recently suggested that the 53 μm OH pump lines in addition to line overlap of large ($\gtrsim$10 km s^{-1}) turbulent line widths can account for the observed dominance of the 1667 MHz transition in OHMs. They further argue that pumping due to FIR radiation can explain all observed main-line OH masers, both those in Galactic star-forming regions and those in OHM galaxies. Given the conditions that exist in ULIRGs and considering that so many OH masers in our Galaxy are associated with massive star-forming regions (Fish et al. 2003), it is therefore not surprising that the entire OHM sample finds homes in LIRGs, strongly favoring the most FIR-luminous, the ULIRGs (Darling & Giovanelli 2002).

The high gas and energy densities in ULIRGs make them natural locations to expect very strong magnetic fields. Much of the radio emission in ULIRGs is resolved on scales of $\sim$100 pc with VLA observations (Condon et al. 1991). High-resolution observations of Arp 220 (Rovilos et al. 2003) show that the OHMs arise in this region as well. With this size scale and the observed radio flux densities, minimum energy arguments suggest *volume-averaged* field strengths of $\approx$1 mG (e.g., Condon et al. 1991; Thompson et al. 2006), which are significantly larger than the $\sim$10 μG fields in normal spirals. The field strengths in ULIRGs cannot be much below a milligauss or else inverse Compton cooling would dominate over synchrotron cooling, making it energetically difficult to explain the radio flux densities from ULIRGs and the fact that ULIRGs lie on the FIR-radio correlation. The field strengths could, however, in principle be larger than the minimum energy estimate if, as in our Galaxy, the magnetic energy density is in approximate equipartition with the total pressure (Thompson et al. 2006). The latter can be estimated from the observed surface density. CO observations of Arp 220 and several other systems reveal $\sim 10^9$ $M_\odot$ of molecular gas in the central ~ 100 pc (e.g., Downes & Solomon 1998), implying gas surface densities $\Sigma \sim 1\text{--}10$ g cm^{-2}, $10^3\text{--}10^4$ times larger than in the Milky Way (MW). The equipartition field scales as $B \propto \Sigma$, implying that the mean field in ULIRGs could approach ~ 10 mG.

Motivated by the above considerations, we carried out a survey of eight (U)LIRGs searching for Zeeman splitting in OHMs. This paper presents our results, which represent the first detections of extragalactic Zeeman splitting from an emission line and the first extragalactic detections within an external galaxy proper. The only previous extragalactic detection was made by Kazes et al. (1991) and confirmed by Sarma et al. (2005) via absorption of 21 cm emission in a high-velocity system toward NGC 1275 (Per A). Section 4.2 outlines what is known about each of our targets. In § 4.3 we describe the observations. In § 4.4 we discuss the data reduction and calibration method. In §§ 4.5 and 4.6 we present a summary and discussion of the results, respectively.

4.2 Source Selection

In selecting our sample of targets from the compilation of all known OHMs by Darling & Giovanelli (2000, 2001, 2002), we chose the three simplest criteria possible. We selected 12 (U)LIRGs: (a) with the largest OHM peak flux densities, (b) whose discoverers did not regard the OHM detection validity as suspicious, and (c) that are observable from Arecibo, Puerto Rico or Green Bank, West Virginia. Our sample includes two of only three known OH gigamasers ($L_{\text{OH}} > 10^4 L_\odot$; Darling & Giovanelli 2002). Here we summarize what is known about each source and its OHM emission.[1]

IRAS F01417+1651.—This LIRG is most commonly known[2] as III Zw 35 and has an optical heliocentric redshift of $z = 0.0274$. It is a double galaxy system and is classified as a Seyfert 2 galaxy. Staveley-Smith et al. (1987) present a single-dish spectrum from the Jodrell Bank Mk1A 76 m telescope showing emission from the 1667 MHz transition at a velocity of 8262 km s^{-1}

---

[1] Unfortunately, we only obtained usable data for two of the six sources observed at Green Bank; therefore, we only provide source descriptions for the eight (U)LIRGs for which we have presentable results.

[2] As a shorthand, when referring to source names in the text by their *IRAS* designation, we henceforth use the right ascension designator only; we refer to 01417, 10038, and 15327 by their more common designators, III Zw 35, IC 2545, and Arp 220, respectively. We retain the full *IRAS* designation in figure and table captions, as well as section headings.

with a peak flux density of 240 mJy and a total velocity extent of 270 km s^{-1} at the 10% flux density level. The line profile can easily be seen to have at least three components. The 1665 MHz line is also weakly detected ($\simeq$25 mJy) and completely separated from the 1667 MHz emission, with an estimated hyperfine line ratio (defined as $R_H \equiv \int f_{1667}\, d\nu / \int f_{1665}\, d\nu$, where the integrals represent the total flux density of each transition) of $R_H \approx 9$.

Killeen et al. (1996) observed III Zw 35 using the Australia Telescope Compact Array (ATCA). Their goal was an attempted detection of Zeeman splitting in the OHM emission. The peak flux density was 247.6 mJy, and their sensitivity was 3.6 mJy. They observed no Zeeman splitting, and their model-dependent estimate for a 3 σ upper limit on the line-of-sight magnetic field was 4.0 mG.

Diamond et al. (1999) present global VLBI observations of the OHM emission in III Zw 35. They label two regions of 1667 MHz OHM emission in the south (S1 and S2) and three in the north (N1–N3), each region covering about 20 mas and separated by 90 mas. They recover 60% of the single-dish flux density. Pihlström et al. (2001) performed simultaneous high-resolution observations of the OHM emission in III Zw 35 using both the European VLBI Network (EVN; baselines between 198 and 2280 km) and the Multi-Element Radio-Linked Interferometer Network (MERLIN; operated by Jodrell Bank Observatory with baselines between 6.2 and 217 km). The map of the 1667 MHz emission shows two compact regions coincident with the northern and southern sources of Diamond et al. (1999) connected by two bridges of weaker, more diffuse emission. In total, 80% of the single-dish flux density was recovered. A velocity gradient of $\simeq$1.5 km s^{-1} pc^{-1} is observed from the southern to the northern regions and is evident in the diffuse component. The emission is modeled as a torus of multiple maser clouds inclined at 60°; the compact OHM emission would be seen at the tangent points where a few clouds could be superimposed in such a fashion that strong OHM emission would be produced from the foreground clouds amplifying those in the background. At the front and back of the torus, the emission would be weak because the path lengths through the torus are small and the clouds are less likely to overlap (Pihlström et al. 2001; Parra et al. 2005).

IRAS F10038−3338.—Also known as IC 2545, this LIRG is a set of interacting galaxies at $z = 0.0341$. A single-dish spectrum made using the Parkes 64 m telescope is presented by Staveley-Smith et al. (1992) showing 1667 MHz OHM emission centered at 10,093 km s^{-1} with a full width at half maximum (FWHM) velocity range of 63 km s^{-1} and a peak flux density of 315 mJy. Likely due to its low declination, there have been no VLBI observations of this source despite its brightness.

Killeen et al. (1996) present an ATCA spectrum with much better sensitivity (5.4 mJy in Stokes I) and velocity resolution than that of Staveley-Smith et al. (1992); the 1667 MHz emission contains five narrow peaks superimposed on a broad emission component. The brightest component has a peak flux density of 260 mJy and a velocity of 10,097 km s^{-1}. Killeen et al. (1996) failed to detect Zeeman splitting and estimated that the line-of-sight field should be less than 4.3 mG.

IRAS F10173+0829.—The only single-dish observations of the OHM line emission in this LIRG at $z = 0.0480$ were made with the 305 m Arecibo telescope and are detailed in Mirabel & Sanders (1987). There are two distinct peaks in the profile of the 1667 MHz emission with a separation of about 100 km s^{-1}, the dominant peak having a velocity of 14,720 km s^{-1} with an FWHM of 39 km s^{-1} and a peak flux density of about 105 mJy. The 1665 and 1667 MHz lines are well separated with a hyperfine line ratio $R_H = 14.6$.

MERLIN observations made by Yu (2004, 2005) show roughly 50 maser spots distributed into three clumps, labeled east, central, and west, over an area of $1\rlap{.}''4 \times 0\rlap{.}''6$. The spots within each clump are distributed along a line, with each of the three lines having a different direction; Yu (2005) proposes (without much justification) that the spots may be distributed along a warped circumnuclear torus seen edge-on. The OHM emission is seen only at the 1667 MHz transition and is coincident with the infrared central position.

IRAS F11506−3851.—Also known as ESO 320-30, this LIRG is classified as an H II galaxy at $z = 0.0108$. A Parkes single-dish spectrum is presented by Staveley-Smith et al. (1992) showing 1667 MHz OHM emission centered at 3103 km s^{-1} with an FWHM velocity extent of 87 km s^{-1} and a peak flux density of 105 mJy. There is neither enough sensitivity nor bandwidth to clearly discern any 1665 MHz emission. There are no interferometric observations of this source: like IC 2545, the low declination of 11506 would hinder any attempted VLBI observations.

IRAS F12032+1707.—A gigamaser discovered at Arecibo by Darling & Giovanelli (2001). The host object, a ULIRG at $z = 0.2170$, has been classified as a LINER-type AGN (Veilleux et al. 1999). The OHM emission spans almost 2000 km s^{-1} with a redshifted high-velocity tail and a mean flux density of roughly 9 mJy. The 1665 and 1667 MHz lines are impossible to distinguish and clearly blended. A very narrow and bright component is seen at 64,500 km s^{-1} with a peak flux density of 16.3 mJy.

Pihlström et al. (2005) used the Very Long Baseline Array (VLBA) to show that the OHM emission is confined to an area of 25×25 mas. All the single-dish flux density was recovered. They were able to clearly identify five peaks in their Stokes I spectrum that corresponded with

Darling's single-dish spectrum. By averaging channels around each peak, they found that the maser components were spatially separated and aligned roughly north-south, implying an ordered velocity gradient. No continuum emission was detected, implying that the continuum emission is resolved out on scales less than 75 mas.

IRAS F12112+0305.—This ULIRG is classified as a LINER-type AGN and is an interacting pair of galaxies at $z = 0.0730$. The only information concerning the OHM emission in this ULIRG is listed in tabular form in Staveley-Smith et al. (1992); no spectrum has been published. The 1667 MHz line was measured at a velocity of 5540 km s^{-1}, and no information about the 1665 MHz transition is published. Its 1667 MHz flux density is listed in Darling & Giovanelli (2002) as 45 mJy. There are no VLBI observations of the 1667 MHz OHM emission for this source.

IRAS F14070+0525.—Discovered by Baan et al. (1992), this gigamaser is the most distant OHM at a redshift of $z = 0.2644$. Darling & Giovanelli (2002) redetected this source in their survey. The OH lines are so wide and blended (1580 km s^{-1} at 10% peak flux density) that it is impossible to identify any 1665 MHz emission. The spectral line profile measured by Darling & Giovanelli (2002) has a peak flux density of 8.4 mJy and shows no significant changes since the original detection by Baan et al. (1992). Darling & Giovanelli (2002) suggest that many peaks in the profile are likely the result of many masing nuclei within the host ULIRG, which is classified as a Seyfert 2.

VLBA observations made by Pihlström et al. (2005) recovered less than 10% of the single-dish flux density, and only two of the many single-dish spectral peaks were detected. Most of the single-dish emission is therefore diffuse. The spatial extent of the VLBA emission is confined to 10×10 mas.

IRAS F15327+2340.—This is perhaps the most well-known ULIRG and is better known as Arp 220 or IC 4553. Baan et al. (1982) discovered the OHM emission using Arecibo and list the single-dish properties of the OHM emission as having a velocity of 5375 km s^{-1} and FWHM velocity extent of 108 km s^{-1}. The spectrum clearly shows that the 1665 and 1667 MHz transitions are distinct with a hyperfine line ratio $R_H = 4.2$ and a peak flux density of 320 mJy.

Smith et al. (1998) used global VLBI continuum imaging at 18 cm to show that the high brightness temperature core of Arp 220 is composed of multiple compact sources, which they interpret as luminous radio supernovae (RSNe). These RSNe are not coincident with the compact 1667 MHz OHM spots discovered by Lonsdale et al. (1998). More recently, Lonsdale et al. (2006) have used high-sensitivity 18 cm VLBI observations of the nuclei to detect four previously unseen sources in a 1 yr period, supporting the RSN interpretation. Parra et al. (2007) have made the first multiwavelength observations of these compact sources; they identify a fraction of these sources to be supernova remnants.

Rovilos et al. (2003) present MERLIN maps of the 18 cm continuum and OHM line emission in Arp 220; two components are seen roughly $1''$ apart, each coinciding with a nucleus imaged in the infrared by Graham et al. (1990). The OHM emission is resolved into one component aligned with the eastern continuum feature and two components that are aligned north to south straddling the western nucleus. Lonsdale et al. (1998) and Rovilos et al. (2003) present global VLBI spectral line maps that show that the OHM emission is resolved into multiple compact spots. The northernmost features in both the eastern and western nuclei form elongated ridges.

4.3 Observations

In 2006 February we used the L-band wide receiver of the 305 m Arecibo[3] telescope in full-Stokes mode in an attempt to detect Zeeman splitting of the 1667 MHz OH transition in the six positive-declination sources listed in § 4.2.

Since the spatial extent of each source is much smaller than Arecibo's $3\overset{.}{.}3$ beam, our observing method was to simply spend equal time at on-source and off-source positions. In this position-switching scheme, we alternated between 4 minutes on source and 4 minutes at a reference position having the same declination as the source and a right ascension 4 minutes east of the source. In this way, the hour angle ranges of the source and reference observations were nearly identical. Our integration time was 1 s, allowing us to remove short-term radio-frequency interference (RFI). The total integration time for each source is as follows: 5.4 hr for III Zw 35, 2.7 hr each for 10173 and 12032, 3.1 hr for 12112, 4.6 hr for 14070, and 5.9 hr for Arp 220. We configured the correlator to produce four spectra per integration: one 6.25 MHz bandpass centered on the mean of the 1665 and 1667 MHz transitions; two narrow bandwidths (either 3.125 or 12.5 MHz, depending on the velocity extent of the source) centered on the 1665 and 1667 MHz transitions, respectively; and one wide bandwidth (either 12.5 or 25 MHz) centered on the mean of the 1665 and 1667 MHz transitions. We calibrated the Mueller matrix for Stokes parameters using the standard Arecibo technique (Heiles et al. 2001b; Heiles & Troland 2004) of observing spider scans on the linearly-polarized continuum sources 3C 138 and 3C 286.

In 2005 December we used the L-band receiver of the 100 m Robert C. Byrd Green Bank

---

[3]The Arecibo Observatory is part of the National Astronomy and Ionosphere Center, which is operated by Cornell University under a cooperative agreement with the National Science Foundation.

Telescope[4] (GBT) to observe an additional six OHM galaxies. All but the two sources in § 4.2 at negative declinations were affected by insidious RFI that left our data corrupted beyond salvation (For completeness, the sources that were obliterated by RFI were IRAS F12540+5708, IRAS F13428+5608, IRAS F17207−0014, and IRAS F20100−4156.) We used two observing methods for each source: position switching (as described above) and least-squares frequency switching (LSFS; for the details of this observing method and its corresponding reduction scheme, see Heiles 2007). The LSFS method was used to accurately derive the gain for each integration; the data were then combined in the standard way using the off-source, position-switched spectra. Our off-source positions were 23 minutes east of each on-source position in order to cause the GBT to track as closely as possible the path of our on-source observations. We used a 12.5 MHz bandwidth and nine-level sampling for all observations.

We accumulated 4.0 hr of RFI-free, on-source integration time for IC 2545 and 5.8 hr for 11506. As we did at Arecibo, we observed 3C 286 using spider scans in order to calibrate the L-band Mueller matrix at the GBT.

4.4 Data Reduction

The complex Stokes I[5] line shape in each of the maser sources is a composite of many narrow maser lines at various velocities spread about the systemic velocity of the system. Therefore, we chose to least-squares fit each line profile with multiple Gaussian components. Without VLBI observations, it is impossible to attribute any particular velocity or width to a Gaussian component within the profile. The only method available to us for assessing a possible field strength from each Stokes V spectrum was to decompose each I profile into the fewest number of Gaussian components that would yield reasonable residuals while also allowing enough components to reproduce the multiple splittings in the V spectrum. None of the parameters in our multiple-component Gaussian fits were held fixed. We discuss our method in more detail in § 4.4.4.

4.4.1 Calibration

The derived Mueller matrix was applied to all OHM observations to correct the polarization products and obtain the pure Stokes spectra for each observed source. We converted from antenna

---

[4]The National Radio Astronomy Observatory is a facility of the National Science Foundation operated under cooperative agreement by Associated Universities, Inc.

[5]We use the classical definition of Stokes I, which is the sum (not the average) of two orthogonal polarizations. Thus, stated Stokes I flux densities are twice those listed in § 4.2 and in other catalogs.

temperature to flux density by assuming the antenna gain to be 10.0 K Jy^{-1} at Arecibo and 2.0 K Jy^{-1} at the GBT; these gains were estimated from observations of standard flux density calibrators.

We follow the IAU definition for Stokes V, namely, $V = \mathrm{RHCP} - \mathrm{LHCP}$, where RHCP is the IEEE definition of right-hand circular polarization.[6] We determined the sense of Stokes V at Arecibo by observing the highly circularly polarized 1665 MHz Galactic maser W49(OH); the result is consistent with the measurements of Rogers et al. (1967) and Coles & Rumsey (1970). The sense of Stokes V has not yet been determined for the GBT Autocorrelation Spectrometer, which only began functioning with full-Stokes capability months prior to these observations.

4.4.2   RFI Removal

We examined each set of spectra, both off source and on, for RFI and rejected suspicious-looking data, which constituted only a few percent for only two sources, III Zw 35 and Arp 220. The other sources were completely free of RFI except for occasional monochromatic signals whose topocentric frequencies are constant; fortunately, most of these fall off the OHM lines.

For one source, 10173, the monochromatic RFI fell on the OHM line. We observed this source over several days, during which the changing Doppler shift moved the RFI across part of the OH spectrum. For each day the RFI was a sharp spike with the usual ringing sidelobes. We Hanning smoothed each spectrum, which eliminated the ringing, and interpolated across each day's spike, which effectively removed the RFI.

The source III Zw 35, whose OH lines are centered near 1622.5 MHz, was highly contaminated by RFI that probably arises from the *Iridium* communications satellites. The RFI consists of a spikey pattern that repeats periodically across the spectrum at about a 0.33 MHz interval. It was impossible to obtain reasonable results by averaging data. However, by taking medians instead of averaging, the spectra look quite good and the RFI is reduced to levels of about 20 mJy in Stokes I and 3 mJy in Stokes V, levels that are considerably smaller ($<1\%$) than the OHM spectral features.

4.4.3   Bandpass and Gain Correction

We correct our spectra for the intermediate-frequency bandpass. Both the Arecibo L-band wide receiver and the GBT L-band receiver are dual-polarized feeds with native linear polarization. For Stokes I and Q, we divide each of the two linear polarization spectra (XX and YY) by its associated off-source spectrum; then we add the results to obtain Stokes I and subtract them to

---

[6]Defined as a clockwise rotation of the electric vector along the direction of propagation.

obtain Q. To generate Stokes U and V, we combine the cross-correlation spectra (XY and YX) having divided by the square root of the product of the off-source XX and YY spectra.

We always show difference spectra: on-source minus off-source. Normally, on- and off-source spectra are combined by subtracting the latter from the former. If the two spectra have equal noise σ, then the noise in the difference is $\sqrt{2}\sigma$. Our off-source spectra have no fine-scale frequency structure, so we can reduce the noise by smoothing. We use a Fourier technique to smooth the off-source average spectrum. By zeroing lags at high delays in the autocorrelation function of the average off-source spectrum and then Fourier transforming, we nearly eliminate the noise contribution from the off-source spectrum while retaining the shape of the bandpass. This reduction in noise is particularly important for the polarized Stokes parameters, which are weak.

As mentioned above, Stokes U and V are obtained via the cross-correlation products XY and YX: this insulates them from system gain fluctuations. However, since Stokes Q is the difference between the two native linear polarizations, it is susceptible to gain fluctuations. We defer the discussion of linear polarization to § 4.5.2.

After gain and bandpass correction, the on- and off-source spectra were averaged separately and combined to yield the final average Stokes spectra.

4.4.4   Fitting Gaussian Components to Stokes I Profiles

Fitting multiple Gaussians to complicated spectral profiles carries a significant degree of subjectivity because the fits are nonlinear. Generally, nonlinear fits require beginning from initial "guessed" parameters and letting the fit converge with successive iterations (Press et al. 1992). Nonlinear fits usually have multiple χ^2 minima, and the particular minimum selected depends on the initial guesses, which in turn depend on the subjective judgement of the person doing the fitting. Therefore, we outline the following guidelines that were used for selecting the initial guesses for Gaussian components in the fits to the Stokes I spectra:

1. For each peak (i.e., local maximum) in I, we included a single Gaussian component whose three parameters (flux density, central frequency, and frequency width) were visually estimated.

2. Many I peaks are distinctly asymmetric. We fitted these asymmetries by including one or two Gaussian components with visually estimated parameters in addition to the central component of guideline 1.

3. The Gaussian components estimated in guidelines 1 and 2 are usually fairly narrow and lie on top of one or two underlying broader lines: core-halo structure. We included one or two broad Gaussian components to represent these broader lines.

4. With all of the above, our goal was to use the fewest number of Gaussian components that would yield reasonable residuals.

It was straightforward to apply the above guidelines to the sources III Zw 35, IC 2545, 11506, 12112, 14070, and even Arp 220, for which we fitted 18 components. The sources 10173 and 12032 are somewhat more complex. In §§ 4.5.1.1–4.5.1.8 we describe how we applied the above selection guidelines when appropriate.

We stress that our Gaussian component representations are not unique. Particular problems include the following:

1. In guideline 1 above, noise prevents us from identifying weak components. This introduces a sensitivity cutoff. Noise also prevents us from distinguishing two or more closely spaced blended real components from a single broader component. Because we favor choices with the fewest number of components, this introduces a bias toward wider components.

2. In guidelines 2 and 3 above, whether to represent a peak needing multiple Gaussians by an asymmetry or core-halo structure can be extremely subjective. For example, the combination of two narrow Gaussians separated by a fraction of their FWHMs can closely mimic the combination of a broad and a narrow Gaussian with roughly the same centers.

In summary, for most sources our Gaussian fits follow our fitting guidelines in a reasonably straightforward fashion; if the fits were done by other people who followed these guidelines, the components would be mostly reproduced. However, Gaussian component fitting has uncertainties as mentioned above, particularly when components are blended and signal-to-noise ratio (S/N) is low.

4.5 Results

For each source, the Stokes I spectrum exhibits a fairly broad, relatively smooth underlying profile for the OHM emission on top of which small bumps from individual masers can be seen. VLBI studies show that the underlying profile arises from spatially extended OH emission and, sometimes, an assembly of many masers that are not individually recognizable (Pihlström et al. 2005; Diamond et al. 1999). We fit the Stokes I spectrum for each source with a series of

Gaussian profiles. For Arp 220, 12112, and 12032 we also fit a first-degree polynomial (12032 required a second-degree polynomial in addition), since these profiles exhibit broad wings. A few sources have a large number of discernible individual masers: for example, we used 18 Gaussian components for Arp 220.

We examined circular polarization for each source and linear polarization for the six Arecibo sources only. Five of the eight sources exhibit significant circular polarization that is interpretable as Zeeman splitting, particularly for the recognizable individual maser components. For four of the sources, there is evidence that the magnetic field reverses direction between OHM spots within the source.

We see detectable linear maser polarization in two sources (possibly four) and are able to estimate Faraday rotation in both. We present all spectra as a function of frequency as viewed in the heliocentric frame. Since all OHMs are extragalactic sources, OHM spectra are almost always presented versus optical heliocentric velocity $v_\odot$, which is conventionally defined as

$$\frac{v_\odot}{c} \equiv \frac{\nu_0}{\nu} - 1 \equiv z_\odot \,, \tag{4.1}$$

where c is the speed of light, ν_0 is the rest frequency (which is taken to be 1667.359 MHz for OHMs since this transition always dominates the 1665.4018 MHz transition), ν is the observed frequency, and $z_\odot$ is the redshift of the maser.

First, in § 4.5.1 we present the circular polarization results for each source in addition to describing the total intensity properties. We present the linear polarization results for each source observed with Arecibo in § 4.5.2. For sources in which the 1665 and 1667 MHz emission lines are separable, we calculate the hyperfine ratio R_H.

4.5.1   Circular Polarization and Line-of-Sight Magnetic Fields

For the usual case in which the Zeeman splitting is small compared to the line width, the Stokes V spectrum is given by

$$V = \left(\frac{\nu}{\nu_0}\right)\left(\frac{dI}{d\nu}\right) bB_\| \,, \tag{4.2}$$

where $B_\|$ is the line-of-sight component of the magnetic field at the OHM and b is known as the *splitting coefficient*[7] (Heiles et al. 1993), equal to 1.96 Hz μG^{-1} for the OH 1667.359 MHz transition;[8] the factor ν/ν_0, equivalent to $(1 + z_\odot)^{-1}$, accounts for the frequency compression of

---

[7]The splitting coefficient is directly proportional to the Landé g-factor for the transition: $b = 2g\mu_0/h$, where μ_0 is the Bohr magneton and h is Planck's constant.

[8]Modjaz et al. (2005) made a valiant effort to detect Zeeman splitting of 22.2 GHz H_2O megamasers in NGC 4258 using the VLA and the GBT, but the splitting coefficient for this hyperfine transition is nearly 1000 times weaker than that of the 1667 MHz OH transition.

redshifted lines. In order to derive a magnetic field strength, we need to least-squares fit the Stokes V spectrum with the functional form of equation (4.2). As is the custom in radio Zeeman work, we add a term on the right that is linear in Stokes I to account for leakage of I into the measured V. For $B_\parallel > 0$ (by convention a positive magnetic field points away from the observer), if Stokes V is plotted as a function of frequency, V will be positive on the low-frequency side of a Stokes I emission line.

We solved equation (4.2) in two ways. In one, we simultaneously fitted Stokes V for multiple Gaussian components (selected as outlined in § 4.4.4) to derive separate, independent magnetic fields for each Gaussian. In the other, we chose a limited range in frequency ν, either 0.1 or 0.25 MHz, and fitted for $B_\parallel$ for the center of this range, positioning the center sequentially at each spectral channel to obtain $B_\parallel$ as a function of frequency; we refer to this as the $B(\nu)$ *fit*. The former method is appropriate for individual masers, while the latter is more suitable for the broad component. We plot the $B(\nu)$ results only for sources for which the results provide additional insight.

Performing these fits requires the calculation of $dI/d\nu$. The I profiles are somewhat noisy and the frequency derivative is often very noisy. This means that traditional least-squares fitting cannot be used because it assumes no error in the independent variables. Sault et al. (1990) discuss this and suggest using a generalized maximum likelihood technique. We choose the much simpler approach of using our multiple-Gaussian fit to the Stokes I spectrum as the independent variable: it has no noise, so it satisfies the requirements of the conventional method of least squares.

We present two vertically stacked plots for each source (Figs. 4.1–4.8). In the top panel, the Stokes I spectrum is plotted as a solid line and the profiles of the Gaussian components are plotted as dashed lines. The residuals (the difference between the data and the composite Gaussian fit) are plotted with enhanced vertical scale as a solid line near the middle height of the panel. Scale bars are plotted on both the residuals and the baseline of the spectrum to the right of the OHM emission; the height of each scale bar corresponds to the labeled flux density. The bottom axes of both plots show heliocentric frequency, and the top axis of the top plot displays the optical heliocentric velocity. In the bottom panel, the Stokes V spectrum is plotted as a solid line and the dashed line represents the best fit to equation (4.2). The integers located between the top and bottom panels label the number of each Gaussian component as assigned in the corresponding tabular summary and are positioned at the central frequencies of each component.[9] The displayed

---

[9]Where multiple labels overlap, the font size has been reduced and the labels stacked corresponding to their associated flux densities.

TABLE 4.1
IRAS F01417+1651 (III ZW 35) GAUSSIAN FIT PARAMETERS

Gaussian (1)	S (mJy) (2)	ν (MHz) (3)	$\Delta\nu$ (MHz) (4)	$v_\odot$ (km s^{-1}) (5)	$B_\parallel$ (mG) (6)
0	190.71 ± 6.77	1622.3743 ± 0.0049	0.2041 ± 0.0081	8312.6	2.94 ± 0.18
1	255.68 ± 22.16	1622.6881 ± 0.0277	0.2786 ± 0.0514	8253.0	-0.47 ± 0.18
2	99.26 ± 6.28	1622.7237 ± 0.0103	0.9694 ± 0.0295	8246.2	1.73 ± 0.78
3	176.13 ± 34.75	1622.7604 ± 0.0020	0.0905 ± 0.0080	8239.3	-2.73 ± 0.13
4	1.17 ± 59.50	1622.8864 ± 0.0142	0.1544 ± 0.0239	8215.3	-3.59 ± 0.26

spectra and residuals are smoothed over seven channels for every source.

We assume that each Gaussian represents an emission component for the 1667 MHz transition. Arp 220 and 12032 have line profiles that are too complex for the 1665 and 1667 MHz lines to be distinguished. This introduces some uncertainty in our Zeeman splitting interpretations in §§ 4.5.1.5 & 4.5.1.8. For sources where the 1665 MHz transition is not blended with the 1667 MHz emission, we present the spectra showing both transitions in § 4.5.2 and calculate the hyperfine line ratios. In all cases the 1665 MHz transition was too weak for Zeeman splitting to be detected even if observed in the 1667 MHz line.

4.5.1.1   IRAS F01417+1651 (III Zw 35)

As we mentioned in § 4.4.2, our observations of III Zw 35 suffered severe RFI that we were able to greatly reduce by combining the data using medians instead of averaging. There remains a spikey pattern that repeats periodically across the spectrum at an interval of $\simeq$0.33 MHz. Remarkably, this spikey pattern is restricted to an 8 MHz wide interval centered almost exactly on the OHM lines. The spikey pattern appears in both the on-source and off-source spectra, so we regard this as terrestrial interference. Despite the RFI, Figure 4.1 shows that both I and V are well detected. In fitting Gaussians we are conservative because we realize that the RFI may have contaminated the line shape. In particular, the 0.33 MHz intervals happen to fall close to the two peaks in Stokes I.

Table 4.1 lists the parameters of the Gaussian components that best fit the Stokes I spectrum. Column (1) lists the zero-based component number. Column (2) lists the peak flux density of each component in mJy and the corresponding uncertainty. Column (3) lists the central heliocentric frequency of each component in MHz and the corresponding uncertainty. Column (4) lists the FWHM of each component in MHz and the corresponding uncertainty. Column (5) lists the

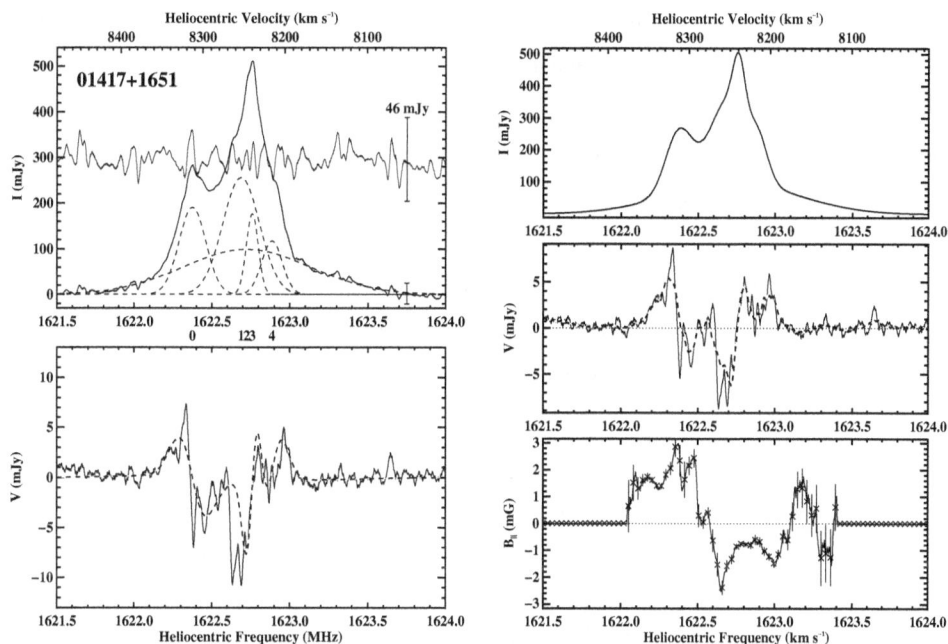

FIG. 4.1—Total intensity and circular polarization results for IRAS F01417+1651 (III Zw 35). *Top left*: Stokes I spectrum (*solid line*; twice the conventionally defined flux density) as a function of heliocentric frequency (*bottom axis*) and optical heliocentric velocity (*top axis*). The profile of each Gaussian component is plotted as a dashed line, with the corresponding component number shown below the frequency axis at the corresponding central frequency. Residuals from the composite Gaussian fit are plotted through the center of the panel (*thin solid line*) and are expanded by a factor of 4. The scale bars near the right edge of the plot correspond to the labeled flux density range. *Bottom left*: Stokes V spectrum (*solid line*) and its fit (*dashed line*). *Top right*: Composite Gaussian fit to Stokes I. *Middle right*: Measured Stokes V (*solid line*) and hypothetical Stokes V (*dashed line*) produced by a uniform $B_{\parallel} = 1$ mG using the derivative of the composite I profile above. *Bottom right*: Derived $B_{\parallel}$ (*crosses and solid line*) and uncertainty (*error bars*) from the $B(\nu)$ fit. All spectra and residuals are smoothed with a boxcar of seven channels.

optical heliocentric velocity corresponding to the central frequency of each component. Column (6) lists the derived line-of-sight magnetic field in mG for each component and the corresponding uncertainty.

Applying our guidelines from § 4.4.4 yielded the five Gaussian components shown in Figure 4.1 and listed in Table 4.1. There are two distinct peaks in the OHM emission. The peak nearest 1622.8 MHz is asymmetric in such a way that two narrow components needed to be added near this peak in order to minimize the residuals. The overall profile has a core-halo structure, with the four narrow components lying on top of a broader component.

The Stokes V spectrum ($S_{\mathrm{rms}} = 0.97$ mJy) in Figure 4.1 shows prominent features that are fitted reasonably well by the five Gaussian components, with Zeeman splitting yielding significant fields in three Gaussians: Gaussian 0 has $B_{\parallel} = 2.9 \pm 0.2$ mG, and Gaussians 3 and 4 have fields of -2.7 ± 0.1 and -3.6 ± 0.3 mG, respectively. Thus, the field reverses from one peak to the other. Pihlström et al. (2001) present 13 spectra from the 1667 MHz OHM emission of III Zw 35 that they mapped using the EVN. These data show clearly that the 8215 and 8240 km s^{-1} components (Gaussians 3 and 4) arise from the southern peak, while the 8312 km s^{-1} component (Gaussian 0) is associated with the northern peak. This provides clear evidence that the reversal is arranged with the magnetic field pointing away from us in the north and toward us in the south.

The right panels of Figure 4.1 show results relevant to the $B(\nu)$ fit. The top panel shows the composite Gaussian-fitted (noise-free) Stokes I spectrum. The middle panel shows the measured Stokes V spectrum as a solid line; the dashed line represents the Stokes V spectrum that would be produced by a uniform line-of-sight magnetic field of 1 mG: this is obtained from equation (4.2) by setting $B_{\parallel} = 1$ mG and using the derivative of the composite Gaussian shown in the top panel. The bottom panel shows the $B(\nu)$ fit (the derived $B_{\parallel}$ as a function of frequency) as described in § 4.5.1. There is a clear systematic pattern, with the field reversing sign from one peak to the other. The estimated field strengths are also consistent with the Gaussian fits.

Parra et al. (2005) and Pihlström et al. (2001) present models of the OHM emission in III Zw 35 as a clumpy, rotating starburst ring at an inclination of 60°, with an inner radius of 22 pc and a radial thickness of 3 pc. Both the Gaussian and $B(\nu)$ analyses suggest that an azimuthal magnetic field is embedded within this starburst ring such that the field points toward us at the southernmost tangent point and away from us at the northernmost tangent point. Parra et al. (2005) estimate that the OHM clouds would be magnetically confined by a magnetic field of order ~ 10 mG.

Killeen et al. (1996) used the ATCA to observe the OHM emission in III Zw 35. Their sensitivity of $S_{\mathrm{rms}} = 3.6$ mJy in Stokes V was not sufficient to detect the Zeeman splitting of the

TABLE 4.2
IRAS F10038−3338 (IC 2545) GAUSSIAN FIT PARAMETERS

Gaussian (1)	S (mJy) (2)	ν (MHz) (3)	$\Delta\nu$ (MHz) (4)	$v_\odot$ (km s^{-1}) (5)	$B_\parallel$ (mG) (6)
0	47.06 ± 3.10	1612.3870 ± 0.0044	0.1505 ± 0.0125	10221.0	1.58 ± 0.75
1	179.73 ± 5.15	1612.8661 ± 0.0032	0.1747 ± 0.0077	10128.9	-1.76 ± 0.26
2	83.85 ± 5.50	1613.0090 ± 0.0119	0.9548 ± 0.0405	10101.4	-11.28 ± 1.16
3	227.46 ± 36.43	1613.0362 ± 0.0052	0.1222 ± 0.0095	10096.2	-0.07 ± 0.18
4	2.85 ± 14.04	1613.1570 ± 0.0089	0.1703 ± 0.0157	10073.0	-0.13 ± 0.16
5	97.85 ± 4.75	1613.3556 ± 0.0035	0.1136 ± 0.0075	10034.9	1.67 ± 0.33

1667 MHz line.

4.5.1.2 IRAS F10038−3338 (IC 2545)

As the residuals in Figure 4.2 show, the 1667 MHz OHM emission from IC 2545 is fitted extremely well by five narrow Gaussian components and one broad one. Using our prescription from § 4.4.4, we see that there are four distinct peaks. The peak nearest 1613.1 MHz has an asymmetry that can be represented with a single extra narrow component. A broad component represents the evident core-halo structure of the OHM emission profile. It is unclear if the emission feature at 1611.7 MHz corresponds to 1665 MHz emission or redshifted 1667 MHz emission. The Stokes I flux density and line profile have not changed since the observations of Killeen et al. (1996). Table 4.2 lists the Gaussian fit parameters shown in Figure 4.4. The Stokes V spectrum has an rms noise of $S_{rms} = 0.7$ mJy. There are three clear detections: Gaussian 1 probes a field of -1.8 ± 0.3 mG, Gaussian 2 is fitted by a field of -11.3 ± 1.2 mG, and Gaussian 5 shows a reversal in sign with a field of 1.7 ± 0.3 mG. Since no VLBI observations exist for this LIRG, nothing can be said about the structure of the field reversal.

4.5.1.3 IRAS F10173+0829

As mentioned in § 4.4.2, we used Hanning-smoothed spectra when least-squares fitting this source because of RFI. We increased the derived uncertainties in Table 4.3 by the appropriate factor of $(8/3)^{1/2}$ (Killeen et al. 1996, Table A1).

We fit Stokes I with seven Gaussians as shown in Figure 4.3 and listed in Table 4.3. Using the selection guidelines of § 4.4.4, we required four narrow components to sufficiently fit the extremely asymmetric peak near 1589.3 MHz. The fit to Stokes I yielded reasonable residuals by

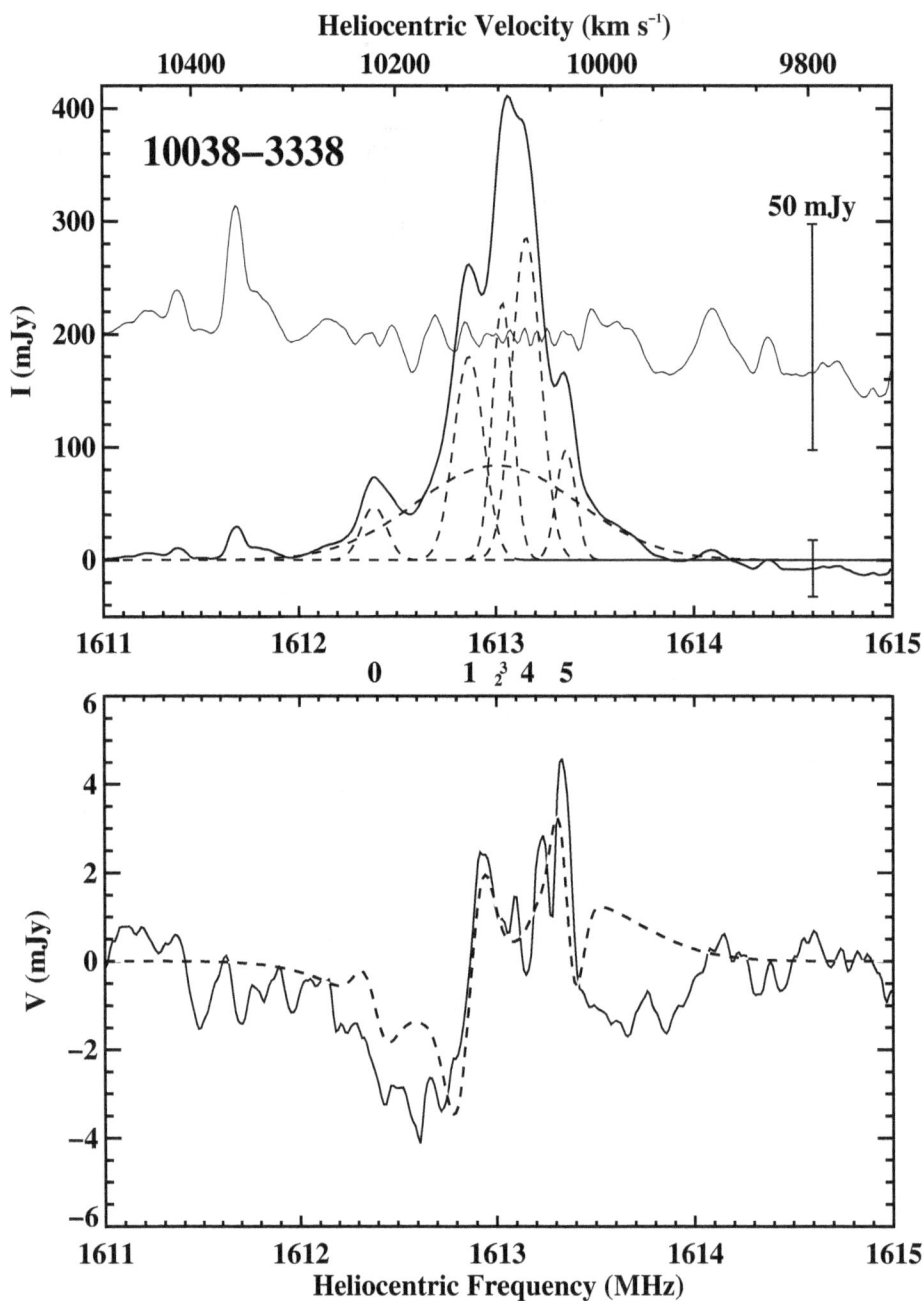

FIG. 4.2 — Total intensity and circular polarization results for IRAS F10038−3338 (IC 2545). See caption for Fig. 4.1. *Top*: Residuals are expanded by a factor of 4.

TABLE 4.3
IRAS F10173+0829 GAUSSIAN FIT PARAMETERS

Gaussian (1)	S (mJy) (2)	ν (MHz) (3)	$\Delta\nu$ (MHz) (4)	$v_\odot$ (km s^{-1}) (5)	$B_\parallel$ (mG) (6)
0	30.62 ± 6.34	1589.2420 ± 0.0037	0.0518 ± 0.0113	14735.9	2.34 ± 2.50
1	16.98 ± 5.95	1589.2862 ± 0.0041	0.0244 ± 0.0106	14727.1	4.19 ± 3.42
2	161.91 ± 7.80	1589.3190 ± 0.0024	0.1786 ± 0.0050	14720.7	0.25 ± 0.89
3	15.54 ± 7.63	1589.3258 ± 0.0073	0.0463 ± 0.0240	14719.3	-2.95 ± 5.15
4	0.42 ± 1.58	1589.5315 ± 0.0194	0.5661 ± 0.0257	14678.6	0.93 ± 5.54
5	3.50 ± 2.82	1589.6383 ± 0.0154	0.0411 ± 0.0412	14657.5	-2.90 ± 17.78
6	16.62 ± 2.46	1589.8772 ± 0.0072	0.1670 ± 0.0279	14610.2	0.80 ± 7.57

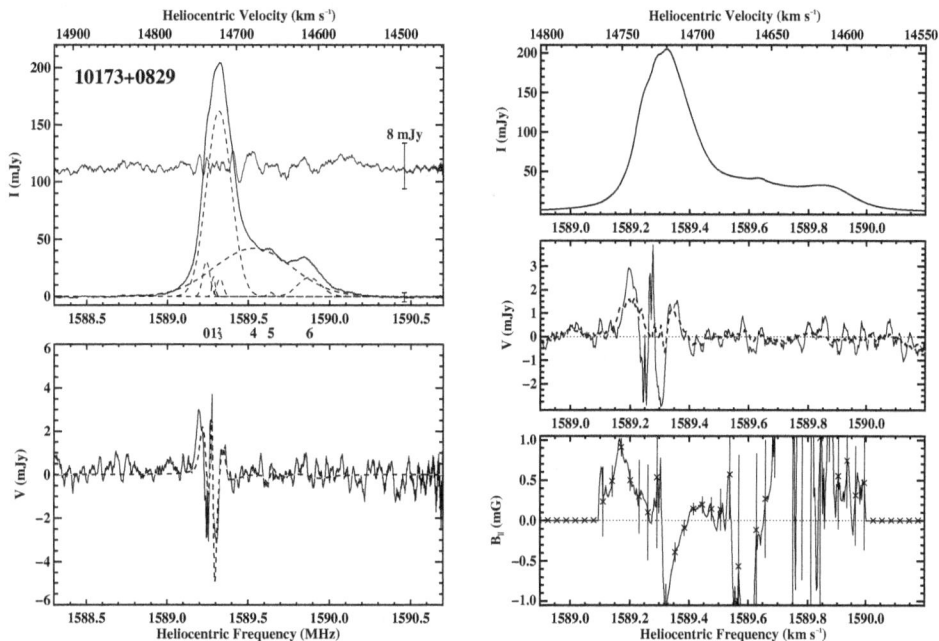

FIG. 4.3 — Total intensity and circular polarization results for IRAS F10173+0829. See caption for Fig. 4.1. *Top left*: Residuals are expanded by a factor of 5. *Middle right*: Dashed line shows the hypothetical Stokes V produced by a uniform $B_\parallel = 2$ mG using the derivative of the composite I profile above.

TABLE 4.4
IRAS F11506−3851 GAUSSIAN FIT PARAMETERS

Gaussian (1)	S (mJy) (2)	ν (MHz) (3)	$\Delta\nu$ (MHz) (4)	$v_\odot$ (km s^{-1}) (5)	$B_\parallel$ (mG) (6)
0	64.73 ± 20.75	1650.0094 ± 0.0428	0.1910 ± 0.0456	3152.3	1.21 ± 0.27
1	102.55 ± 26.92	1650.1891 ± 0.0240	0.1810 ± 0.1018	3119.3	0.45 ± 0.20
2	86.37 ± 83.80	1650.3232 ± 0.0231	0.1358 ± 0.0467	3094.7	0.36 ± 0.17
3	116.30 ± 24.97	1650.4685 ± 0.0257	0.2693 ± 0.0976	3068.0	0.73 ± 0.20
4	0.31 ± 7.25	1650.7843 ± 0.1171	0.3616 ± 0.2594	3010.1	0.68 ± 0.75
5	23.79 ± 7.46	1651.0731 ± 0.0102	0.1260 ± 0.0368	2957.1	1.03 ± 0.50

including a single extremely broad component. This source represents a case where the profile structure is too complex to be modeled by our straightforward fitting guidelines: we regard the derived components as highly suspect regardless of the quality of the fit.

Detectable portions of the Stokes V spectrum are restricted to the strongly peaked I line, where V oscillates rapidly. Due to the asymmetry in this profile, we found it impossible to obtain a set of Gaussians that gives good fits to both I and V. Our final fit reflects a compromise between large residuals and a larger number of Gaussians. While it is clear that Zeeman splitting has been detected in this source, we would need to add many more narrow, weak Gaussian components to our Stokes I fit in order to obtain statistically significant magnetic field derivations from the fit to Stokes V (which has an rms noise of $S_{\mathrm{rms}} = 0.8$ mJy). Without any VLBI observations or any other physical motivation for adding such components, the best we can do is present the evidence for Zeeman splitting without estimating field strengths for individual maser components.

The right panels of Figure 4.3 show plots relevant to the $B(\nu)$ fit. The dashed line in the middle panel shows the Stokes V spectrum that would be expected for a uniform $B_\parallel = 2$ mG. The bottom panel shows the $B(\nu)$ fit. There is a clear systematic pattern, with the field reversing sign from one side of the peak to the other. This reversal is not revealed by the Gaussian fits because there are no narrow Gaussians on either side of the peak and because the peak itself is represented by a single Gaussian (component 2).

Since no spectral information is presented in the MERLIN maps of Yu (2004, 2005), it is impossible to associate any of our Gaussian components with OHM spots in 10173.

4.5.1.4   IRAS F11506−3851

The top panel of Figure 4.4 shows that the 1667 MHz Stokes I emission is fitted quite well by six narrow Gaussian components. The parameters for each component are listed in Table 4.4.

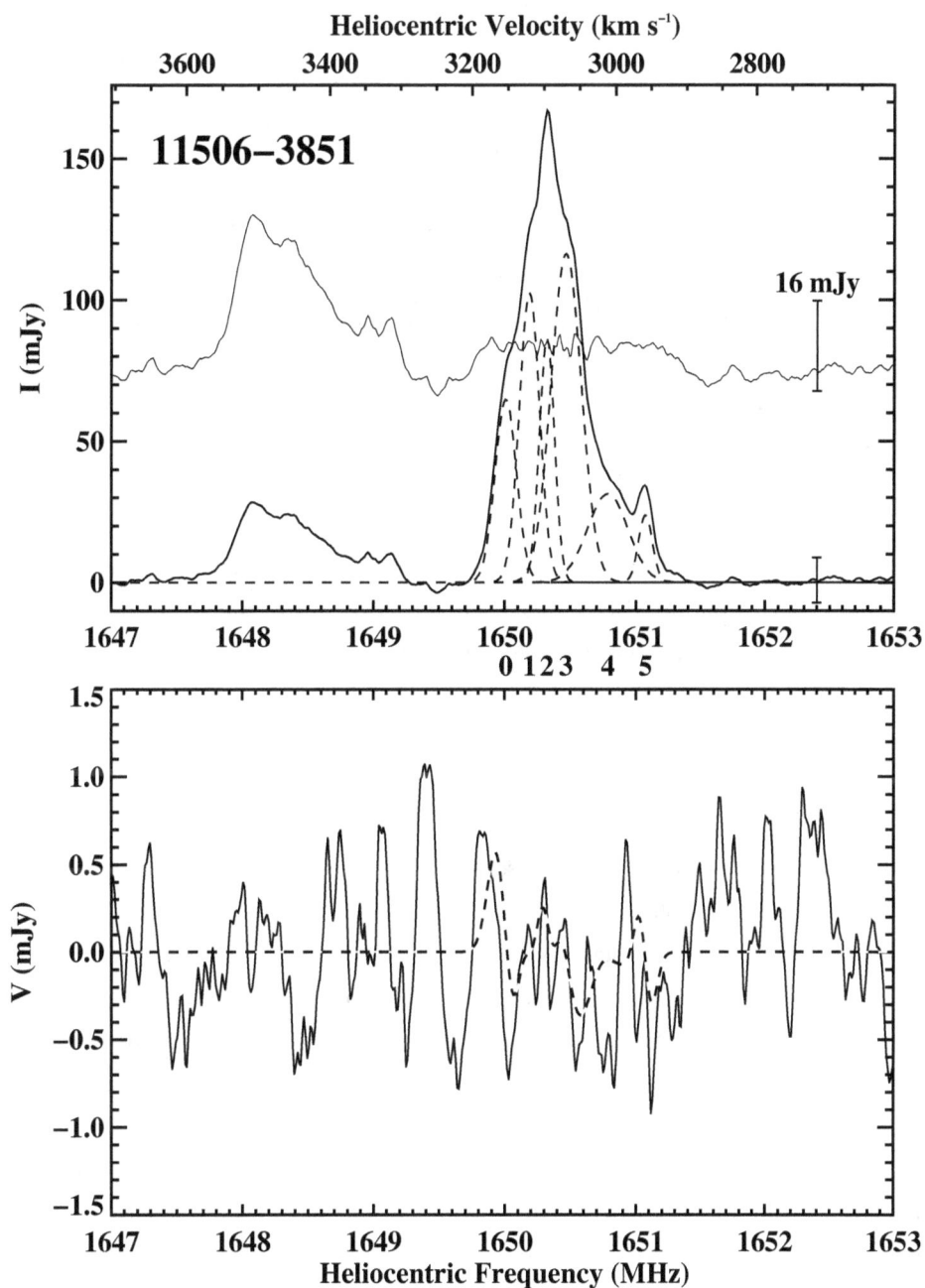

FIG. 4.4 — Total intensity and circular polarization results for IRAS F11506−3851. See caption for Fig. 4.1. *Top*: Residuals are expanded by a factor of 2.

TABLE 4.5

IRAS F12032+1707 GAUSSIAN FIT PARAMETERS

Gaussian (1)	S (mJy) (2)	ν (MHz) (3)	$\Delta\nu$ (MHz) (4)	$v_\odot$ (km s^{-1}) (5)	$B_\parallel$ (mG) (6)
0	4.31 ± 0.54	1367.1002 ± 0.0169	0.4082 ± 0.0370	65844.0	-6.79 ± 7.94
1	3.95 ± 0.96	1367.6739 ± 0.0457	0.7915 ± 0.1793	65690.6	30.42 ± 12.43
2	8.42 ± 0.99	1368.8016 ± 0.0094	0.6264 ± 0.0448	65389.5	6.91 ± 4.86
3	3.00 ± 0.34	1369.5922 ± 0.0071	0.1460 ± 0.0207	65178.7	-1.05 ± 6.48
4	0.21 ± 0.36	1369.6903 ± 0.0452	2.6434 ± 0.1574	65152.6	-3.25 ± 3.98
5	3.23 ± 0.54	1369.9216 ± 0.0036	0.0450 ± 0.0089	65091.0	3.10 ± 3.41
6	6.67 ± 0.34	1370.7194 ± 0.0064	0.2954 ± 0.0165	64878.6	11.64 ± 4.31
7	21.67 ± 0.65	1371.2516 ± 0.0030	0.4769 ± 0.0115	64737.0	10.90 ± 1.72
8	16.99 ± 0.46	1371.3316 ± 0.0010	0.0849 ± 0.0029	64715.8	17.92 ± 0.89
9	16.19 ± 3.38	1372.1576 ± 0.0568	0.4872 ± 0.0531	64496.4	1.78 ± 2.48
10	25.02 ± 4.51	1372.3299 ± 0.0027	0.2795 ± 0.0165	64450.6	-1.45 ± 1.12
11	9.56 ± 0.51	1372.7780 ± 0.0412	0.7315 ± 0.0588	64331.7	-11.69 ± 4.98
12	2.02 ± 0.21	1373.8371 ± 0.0179	0.3807 ± 0.0500	64051.0	20.61 ± 15.49

Gaussian components 0 and 3 have derived magnetic fields that look significant when judged by their formal errors. However, given the quality of the Stokes V spectrum ($S_{\rm rms} = 0.4$ mJy), we have no confidence in either result. The 1665 MHz emission is clearly completely separated from the 1667 MHz emission, yielding a hyperfine ratio $R_H = 4.9$.

4.5.1.5 IRAS F12032+1707

The OHM emission from this gigamaser has an extremely wide extent; it is impossible to distinguish the 1665 and 1667 MHz emission. We fit its Stokes I spectrum with 13 Gaussians as shown in Figure 4.5 with parameters as listed in Table 4.5. Due to the complexity of this profile, this is an extremely difficult case to apply our component selection guidelines to. We first identify eight local maxima marked in Figure 4.5 by components 0, 2, 3, 5, 6, 8, 10, and 12. The peaks near 8 and 10 display shoulders and are clearly blended with narrower components: we added one additional component to each (components 7 and 9, respectively). We represent a broad shoulder, not quite intense enough to produce a local maximum, near 1373 MHz by component 11. Finally, we add two broad halo components, numbers 1 and 4, to minimize the residuals. While the prescription of § 4.4.4 allows us to select these 13 components somewhat straightforwardly, even producing respectable residuals, visual inspection of Figure 4.5 suggests that our model simplifies and glosses over the innate complexity of this OHM profile. Without VLBI observations, there is little that can be done to improve on our model.

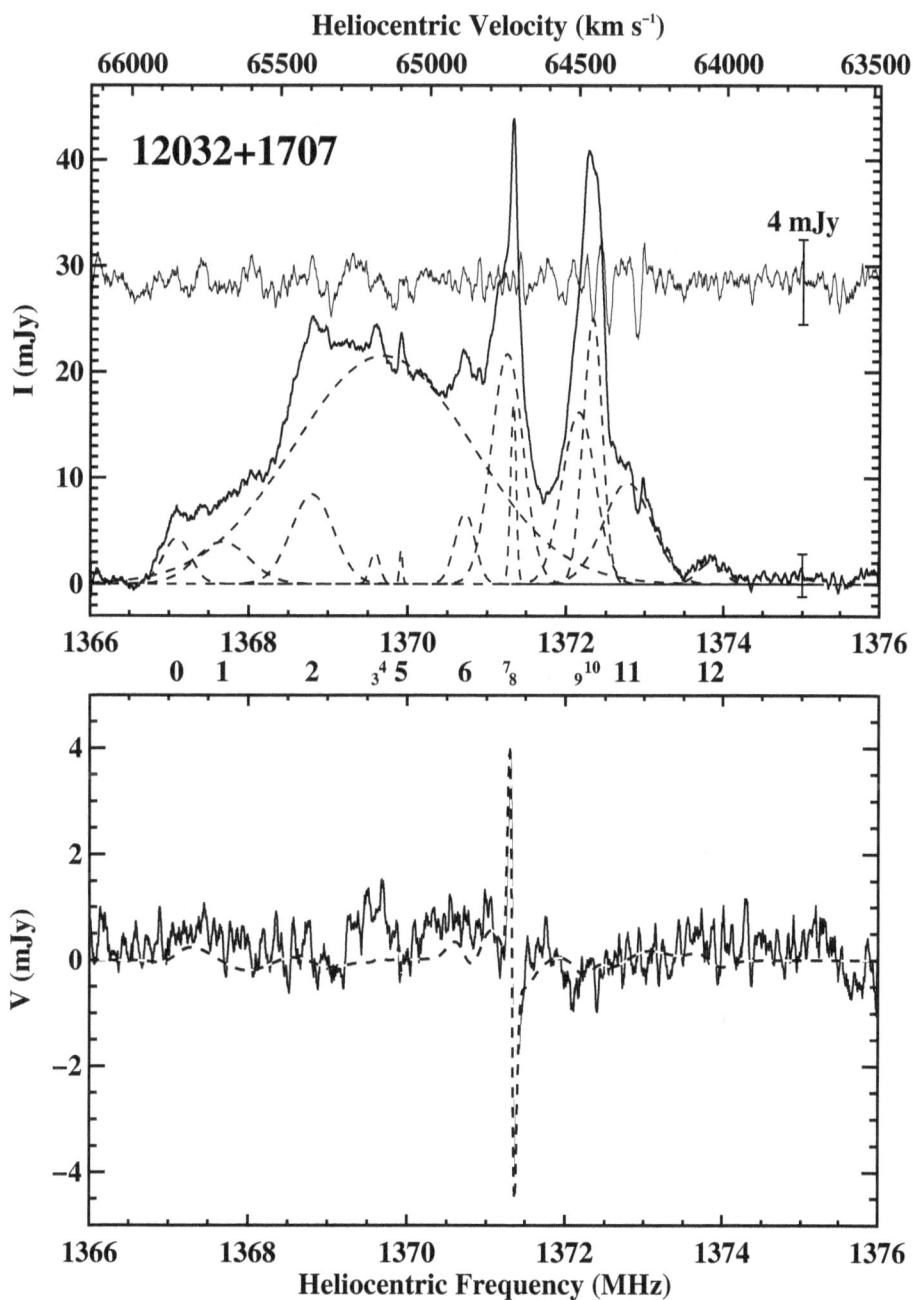

FIG. 4.5—Total intensity and circular polarization results for IRAS F12032+1707. See caption for Fig. 4.1. *Top*: Residuals are expanded by a factor of 2.

TABLE 4.6
IRAS F12112+0305 GAUSSIAN FIT PARAMETERS

Gaussian (1)	S (mJy) (2)	ν (MHz) (3)	$\Delta\nu$ (MHz) (4)	$v_\odot$ (km s^{-1}) (5)	$B_\parallel$ (mG) (6)
0	16.19 ± 0.75	1554.1822 ± 0.0013	0.1467 ± 0.0054	21831.1	-0.20 ± 1.99
1	58.16 ± 0.51	1554.2767 ± 0.0048	0.5188 ± 0.0098	21811.6	0.25 ± 1.17
2	48.63 ± 0.97	1554.5634 ± 0.0007	0.1893 ± 0.0028	21752.3	1.02 ± 0.77
3	13.83 ± 0.59	1554.5641 ± 0.0006	0.0309 ± 0.0017	21752.1	0.27 ± 1.03
4	0.32 ± 0.34	1554.8544 ± 0.0037	0.4359 ± 0.0063	21692.1	3.09 ± 1.92

Detectable signals in the V spectrum ($S_{\rm rms} = 0.9$ mJy) are restricted to the stronger of the two narrow peaks. This narrow peak is somewhat asymmetric and requires two Gaussians for a good fit. These two components (numbers 7 and 8) have field strengths of 10.9 ± 1.7 and 17.9 ± 0.9 mG, respectively. Three other Gaussians, numbers 1, 6, and 11, have fields that are nearly 3 σ; however, visual examination of the V spectrum shows bumps and wiggles throughout at the ≈ 1 mJy level, and we have no confidence in these purported fields.

Since the significant splittings occur where the Stokes I profile has the highest flux density, and since the 1667 MHz transition dominates in all OHMs, we feel comfortable assuming that the emission is from the 1667 MHz transition. Of course, because of the blending of the 1665 and 1667 MHz lines, there is an unresolvable ambiguity that could affect the derived field strengths.

The Stokes I line profile has changed since the source's discovery by Darling & Giovanelli (2001). The flux densities of the broad component and the narrow component near 1372.3 MHz have remained the same. However, the narrow component near 1371.3 MHz, which used to be roughly 6 mJy (using our classical definition of Stokes I) weaker than their 32.5 mJy peak at 1372.3 MHz, has flared and is now the strongest component with a flux density of 44 mJy. This time-variable component is the same one that exhibits the largest splitting and therefore probes the strongest field; in § 4.6 we compare this result with newly observed strong field detections in time-variable Galactic OH maser components.

4.5.1.6   IRAS F12112+0305

We were able to fit the Stokes I OHM emission quite nicely with five Gaussians as shown in Figure 4.6. The fit parameters are listed in Table 4.6. There is no detectable signal in Stokes V ($S_{\rm rms} = 1.2$ mJy). This is the first published spectrum of OHM emission in 12112.

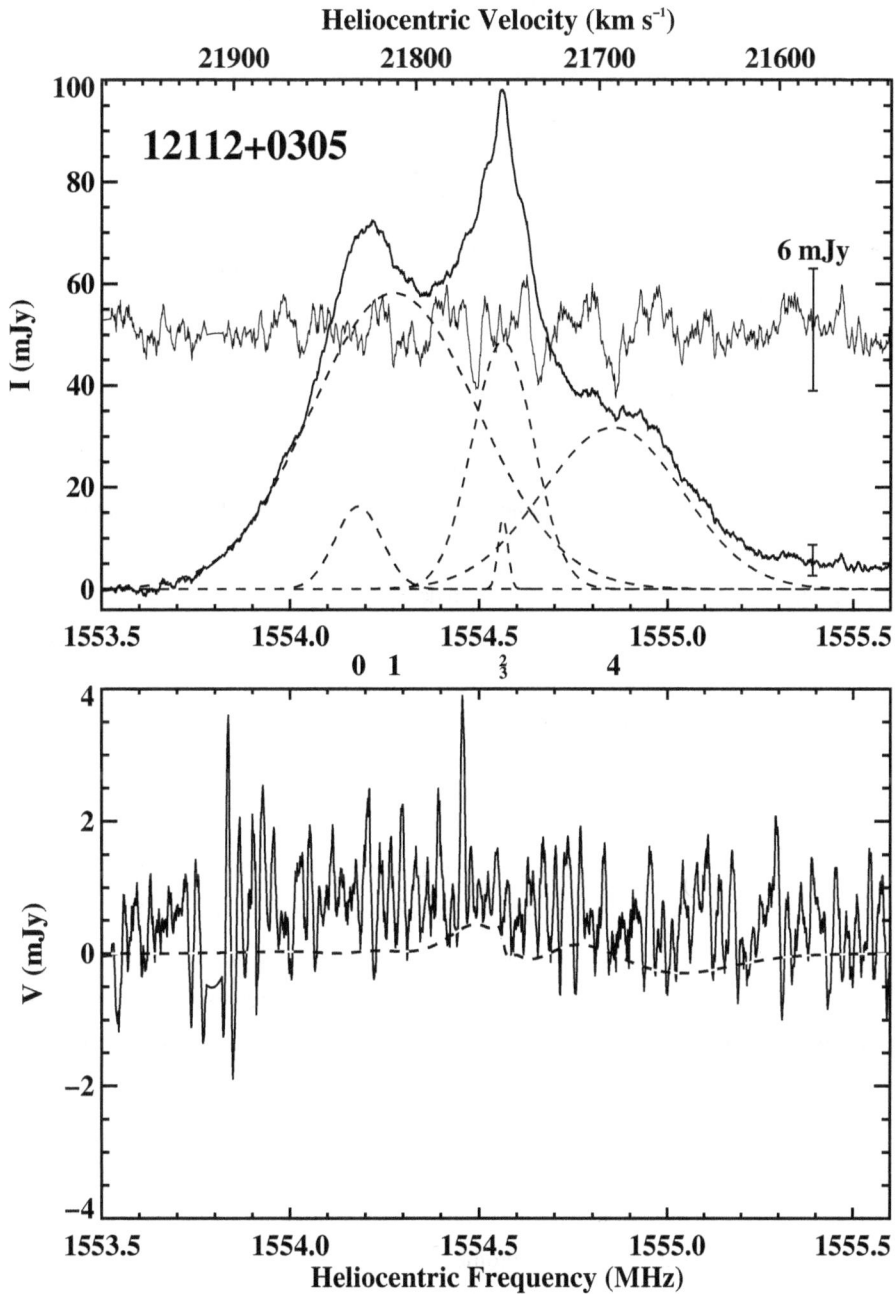

FIG. 4.6—Total intensity and circular polarization results for IRAS F12112+0305. See caption for Fig. 4.1. *Top*: Residuals are expanded by a factor of 4.

TABLE 4.7
IRAS F14070+0525 GAUSSIAN FIT PARAMETERS

Gaussian (1)	S (mJy) (2)	ν (MHz) (3)	$\Delta\nu$ (MHz) (4)	$v_\odot$ (km s^{-1}) (5)	$B_\parallel$ (mG) (6)
0	3.50 ± 0.22	1315.2359 ± 0.0105	0.4088 ± 0.0317	80262.3	-10.42 ± 10.49
1	8.95 ± 0.19	1316.0759 ± 0.0715	2.1625 ± 0.0861	80019.7	26.65 ± 9.57
2	5.70 ± 0.30	1316.2886 ± 0.0047	0.2040 ± 0.0134	79958.4	1.76 ± 4.65
3	11.26 ± 0.68	1316.9061 ± 0.0100	0.6771 ± 0.0302	79780.3	-3.31 ± 4.67
4	0.08 ± 0.50	1317.5519 ± 0.0118	0.4878 ± 0.0271	79594.2	-3.16 ± 5.59
5	2.78 ± 0.26	1318.8802 ± 0.0106	0.2664 ± 0.0307	79212.1	12.72 ± 10.82
6	10.76 ± 0.16	1319.1605 ± 0.0113	1.2737 ± 0.0199	79131.6	-12.92 ± 6.16

4.5.1.7   IRAS F14070+0525

Table 4.7 lists the parameters for the seven Gaussian components used to fit the Stokes I OHM emission in 14070. Since the 1665 and 1667 MHz lines are clearly blended in this source, we assume that all of the components represent 1667 MHz emission. As seen in Figure 4.7, this decomposition provides a decent fit, but there is no detectable signal in Stokes V ($S_{\rm rms} = 0.5$ mJy). Gaussian number 1 shows a nearly 3 σ detection of magnetic field; however, the associated feature in the Stokes V spectrum appears to be no more significant than the other features of millijansky-strength intensity. We have no confidence in this near detection.

4.5.1.8   IRAS F15327+2340 (Arp 220)

The line profile of the Stokes I OHM emission in Arp 220 is very complex. Our fit required 18 Gaussian components, as seen in Figure 4.8, to obtain reasonable residuals. An 18-component fit might seem overwhelming, but the components are easily obtained using our selection guidelines from § 4.4.4. There are nine distinct peaks (i.e., local maxima), represented in Figure 4.8 by Gaussian components 0, 4, 5, 7, 9, 13, 14, 16, and 17. There are six narrow or fairly narrow bumps or shoulders that are not intense enough to produce local maxima, represented by Gaussian components 1, 2, 3, 11, 12, and 15. Component 10 was needed to represent the asymmetry in the brightest peak near 1638.15 MHz. Finally, components 6 and 8 were needed to represent core-halo structure in the overall profile. This 18-component fit reproduces all of the visually obvious narrow, weak bumps, as well as the overall profile shape. However, the residuals exhibit a different signature in the line from that off the line, which means that our fit does not represent the I profile perfectly. We expended considerable effort making sure that each of the 18 Gaussians listed in Table 4.8 is actually needed for the fit by inspecting the residuals for different combinations of

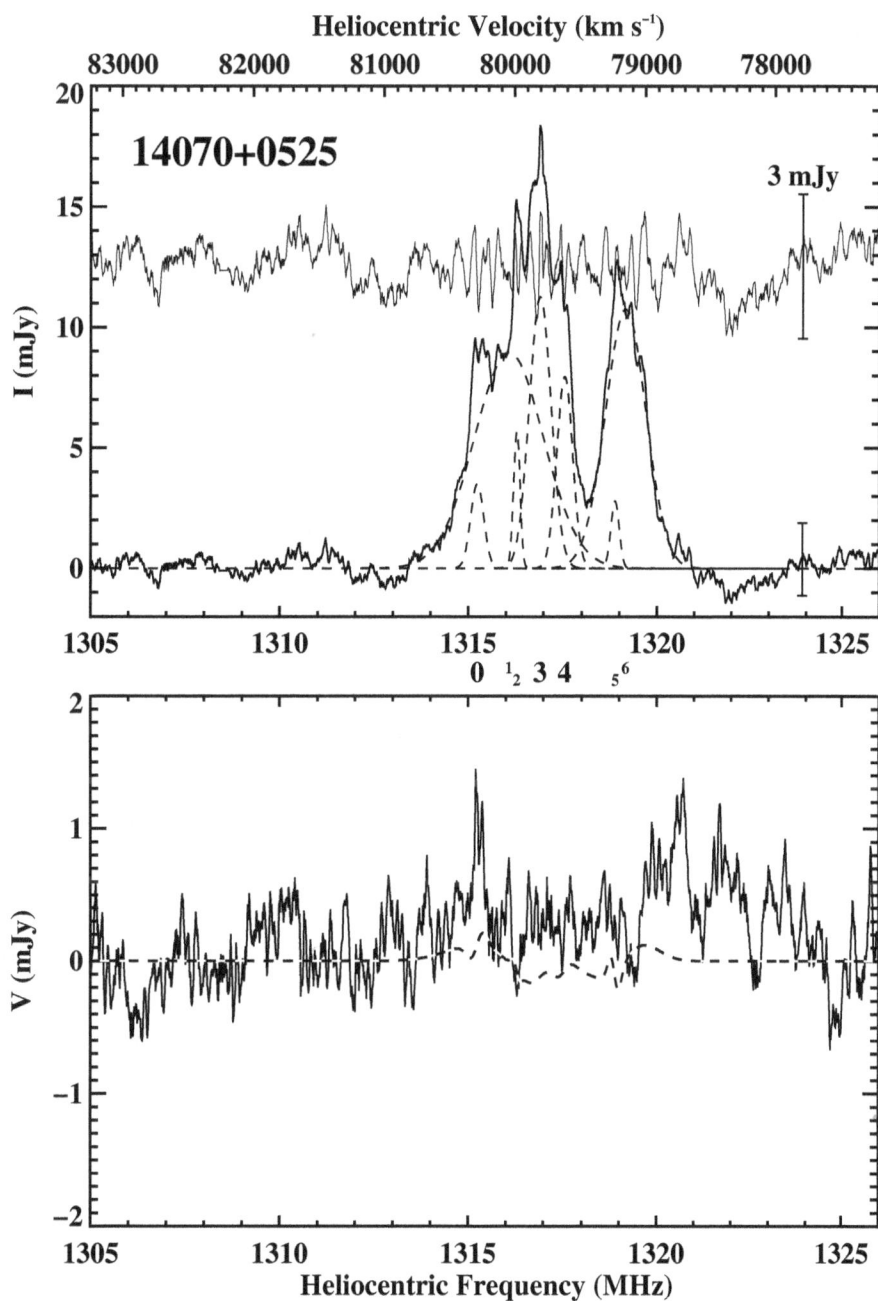

FIG. 4.7—Total intensity and circular polarization results for IRAS F14070+0525. See caption for Fig. 4.1. *Top*: Residuals are expanded by a factor of 2.

TABLE 4.8

IRAS F15327+2340 (ARP 220) GAUSSIAN FIT PARAMETERS

Gaussian (1)	S (mJy) (2)	ν (MHz) (3)	$\Delta\nu$ (MHz) (4)	$v_\odot$ (km s^{-1}) (5)	$B_\parallel$ (mG) (6)
0	14.52 ± 0.98	1637.1424 ± 0.0008	0.0243 ± 0.0020	5533.2	-4.78 ± 0.53
1	10.92 ± 0.61	1637.3105 ± 0.0018	0.0709 ± 0.0050	5501.9	-0.11 ± 1.21
2	11.80 ± 1.02	1637.5074 ± 0.0009	0.0231 ± 0.0024	5465.2	-2.78 ± 0.64
3	10.12 ± 1.00	1637.5736 ± 0.0011	0.0241 ± 0.0029	5452.9	7.77 ± 0.76
4	0.90 ± 0.77	1637.7198 ± 0.0002	0.0590 ± 0.0006	5425.6	-2.78 ± 0.13
5	51.49 ± 1.06	1637.8723 ± 0.0005	0.0621 ± 0.0016	5397.2	0.33 ± 0.25
6	324.77 ± 3.96	1637.8916 ± 0.0010	0.3362 ± 0.0028	5393.6	0.26 ± 0.12
7	97.84 ± 2.60	1638.0196 ± 0.0008	0.0800 ± 0.0023	5369.7	-0.15 ± 0.18
8	293.07 ± 4.06	1638.0313 ± 0.0011	1.0064 ± 0.0075	5367.6	0.14 ± 0.21
9	386.20 ± 5.88	1638.1189 ± 0.0007	0.0882 ± 0.0012	5351.2	-0.76 ± 0.06
10	81.13 ± 2.84	1638.1375 ± 0.0002	0.0292 ± 0.0008	5347.8	-0.24 ± 0.11
11	237.99 ± 4.75	1638.2098 ± 0.0011	0.1039 ± 0.0024	5334.3	0.66 ± 0.10
12	46.95 ± 2.66	1638.3468 ± 0.0051	0.1894 ± 0.0103	5308.8	-1.03 ± 0.59
13	18.22 ± 1.02	1638.4199 ± 0.0008	0.0342 ± 0.0024	5295.2	0.20 ± 0.50
14	19.79 ± 0.78	1638.6066 ± 0.0011	0.0769 ± 0.0036	5260.4	0.22 ± 0.69
15	7.62 ± 0.60	1638.8445 ± 0.0033	0.1038 ± 0.0103	5216.1	1.78 ± 2.10
16	11.23 ± 0.82	1638.9714 ± 0.0014	0.0367 ± 0.0034	5192.5	1.42 ± 0.86
17	13.18 ± 0.57	1639.0663 ± 0.0018	0.0908 ± 0.0051	5174.9	-0.46 ± 1.15

omitted Gaussians.

Six narrow Gaussians exhibit visually obvious signatures in Stokes V (which has an rms noise of $S_{\rm rms} = 1.16$ mJy) and provide good fits for Zeeman splitting. The absolute values of the derived field strengths range from 0.7 to 4.7 mG, with four negative and two positive fields. Gaussians 2 and 3 have opposite field strengths.

Gaussians 9 and 11 are strong (a few hundred millijanskys), have comparable FWHMs of about 0.1 MHz, and are separated by about the FWHM. This makes them easily distinguishable. They have opposite field directions as given by the least-squares fit, but the reversal in sign is also visually apparent. Zeeman splitting produces a Stokes V pattern that looks like the frequency derivative of the line, with amplitude and sign scaled by $B_\parallel$. Thus, for a single Gaussian component, the V pattern looks like the letter "S" lying on its side, with an inevitable negative and positive part; the integral over Stokes V must be zero. However, the V pattern for these Gaussians in Figure 4.8 does not look like this; instead it is positive on both sides of the line and negative in the middle. The only way to obtain positive V on each side of a spectral bump is for the field to have different signs on the two sides (e.g., Verschuur 1969a, Fig. 3, the first radio detection of Zeeman splitting). The integral of Stokes V must again be zero, with the central negative portion balanced by the two positive ones on the sides. The reversed field not only is a result of the fits but

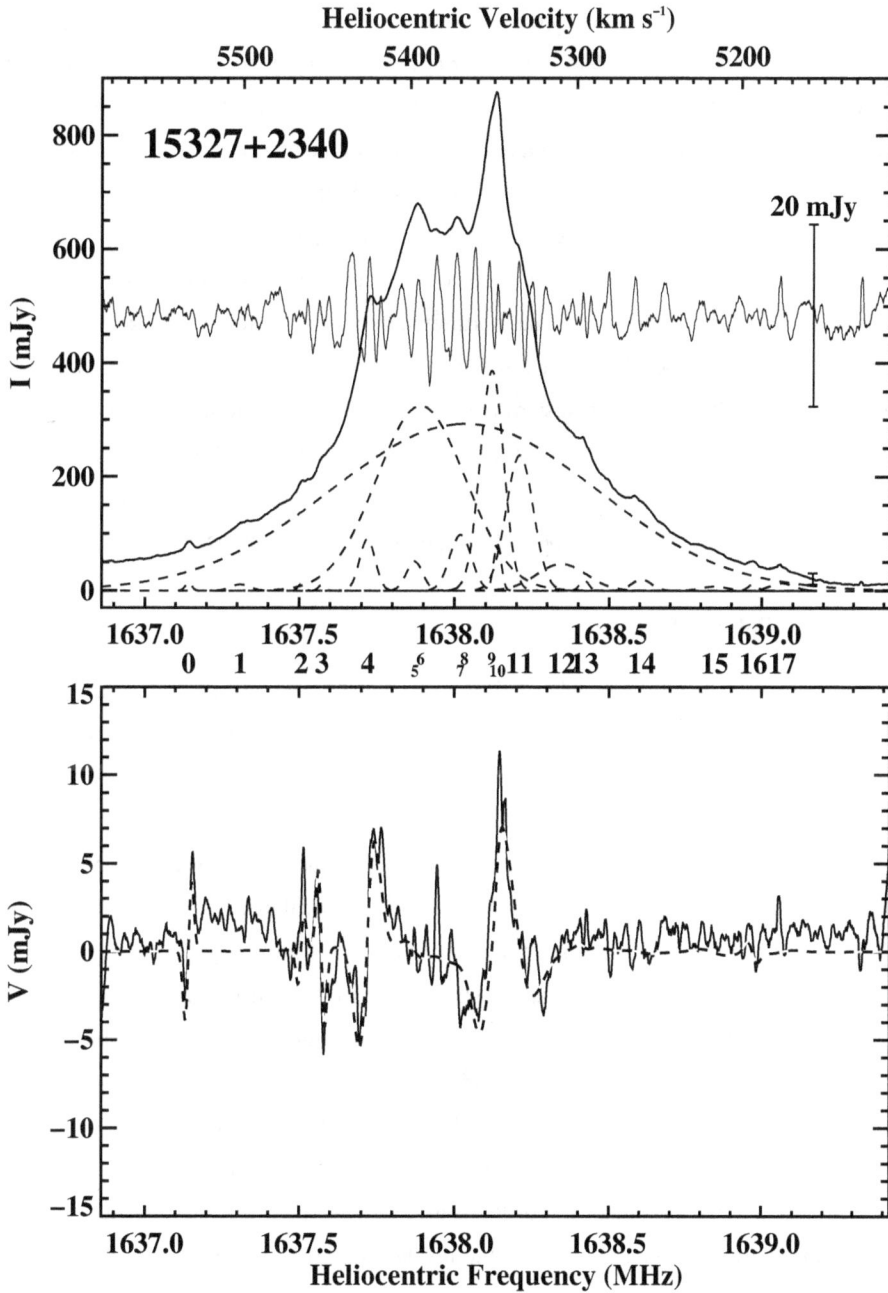

FIG. 4.8 — Total intensity and circular polarization results for IRAS F15327+2340 (Arp 220). See caption for Fig. 4.1. *Top*: Residuals are expanded by a factor of 16.

also is visually apparent.

We can compare our single-dish spectrum with the selected global VLBI spectra presented by Rovilos et al. (2003) and Lonsdale et al. (1998). There are a number of Gaussian components that appear to be directly associated with the resolved OHM spots: component 11 at 5334 km s^{-1} originates in a southwestern spot tracing a positive field; component 6 at 5393 km s^{-1} originates in the southeast and traces a positive field; component 4 at 5425 km s^{-1} originates in the center of the northeast ridge and traces a negative field; component 0 at 5533 km s^{-1} originates in one of the southwestern spots tracing a negative field. Three other features are more ambiguous: components 2 and 3 could be associated with either the northwestern or northeastern OHM ridges, while the brightest component, number 9 at 5351 km s^{-1}, appears to contain emission from both of the northern ridges as well as the southwestern maser spots; these ambiguities prevent any possible field associations. The picture painted by the possible associations is for a field reversal from positive to negative from the southern to the northern features of the eastern OHM spots; there is no obvious reversal in the western region, but it is possible given the associations above.

4.5.2 Linear Polarization

For all observations, we used dual-polarized feeds with native linear polarization. This means that the observed Stokes U_{obs} and V_{obs} come from cross-correlation products, which insulates them from system gain fluctuations. However, Stokes Q_{obs} comes from the difference between the two native linear polarizations, so it is susceptible to time-variable, unpredictable gain fluctuations. This leads to coupling between Stokes I and Stokes Q_{obs}; in other words, a scaled replica of the I profile appears in the Q_{obs} profile, with a random and unknown scaling factor, so Q_{obs} is unreliable.

Normally, when deriving linear polarization, one combines Stokes Q_{obs} and U_{obs} in the standard ways to obtain polarized intensity and position angle for the astronomical source. However, since Q_{obs} is unreliable for our measurements, we derived Stokes Q_{src} and U_{src} for the source from U_{obs} alone by least-squares fitting its variation with parallactic angle. This is quite feasible at Arecibo because all sources pass within $20°$ of the zenith, so tracking for a reasonably long time provides a wide spread in parallactic angle. This makes the least-squares fit robust and provides good sensitivity and low systematics. For the source III Zw 35, which was plagued by serious interference, we performed a minimum-absolute-residual-sum (MARS) fit.

As with the Stokes I and V spectra, the least-squares derived Q_{src} and U_{src} spectra are displayed after subtracting both the off-source position and the continuum. We then use these baseline-subtracted Stokes spectra to derive spectra for polarized intensity and position angle. We do this because, even for the position-switched spectra, the continuum linear polarization is usually dominated by the diffuse Galactic synchrotron background. Although this prevents us from reliably deriving linear polarization for the (U)LIRG continuum radiation, the frequency-variable polarization is reliable.

For two sources below, we least-squares fit for the Faraday rotation measure RM. Performing this fit requires some care because the RM is derived from the position angle ψ, which in turn is obtained by combining Q_{src} and U_{src}, which combine nonlinearly through the arctan function [$\psi = 0.5 \arctan(Q_{\mathrm{src}}/U_{\mathrm{src}})$]. The channel-by-channel data are too noisy to produce a good-looking spectral plot of ψ, so on our plots we boxcar smooth by an appropriate number of points. One cannot linearly fit the unsmoothed values of ψ to frequency because the arctan function produces nonlinear noise in ψ. To avoid this problem, we performed a nonlinear fit to the unsmoothed $\arctan(Q_{\mathrm{src}}/U_{\mathrm{src}})$; this extra complication ensures that the derived values and errors are unaffected by smoothing.

We were unable to analyze the linear polarization for the two sources we observed using the GBT (IC 2545 and 11506) because of inadequate parallactic angle coverage. We report the linear polarization for the Arecibo results here and discuss their interpretation in § 4.6.3.

We present three vertically stacked plots for each source below (Figs. 4.9–4.14). The top panels show the position-differenced, baseline-subtracted Stokes I spectrum over 12.5 MHz, therefore including both the 1665 and 1667 MHz transitions for each source. The middle panels present the linear polarization intensity, and the bottom panels display the derived position angle ψ as a function of heliocentric frequency.

4.5.2.1 IRAS F01417+1651 (III Zw 35)

The linear polarization results for III Zw 35 are presented in Figure 4.9. The top panels exhibit the position-differenced, baseline-subtracted Stokes I spectrum. The 1665 MHz transition is clearly visible, and the hyperfine line ratio is $R_H = 6.0$. The middle panels clearly display that the spectrum of linearly polarized intensity is extremely spikey. Although higher S/N would help, the spikes might be real and possibly correspond to individual masers that are too weak to be seen clearly in the Stokes I spectrum. The polarized intensity shows a seemingly real peak centered near 1622.8 MHz, which is also the center of the Stokes I peak. The polarized intensity is about 5 mJy and the Stokes I peak is roughly 500 mJy, so the fractional polarization is $\approx 1\%$. If the other spikes are real, then their fractional polarizations are much higher.

The bottom panels display the position angle ψ. Position angles exhibit less scatter than

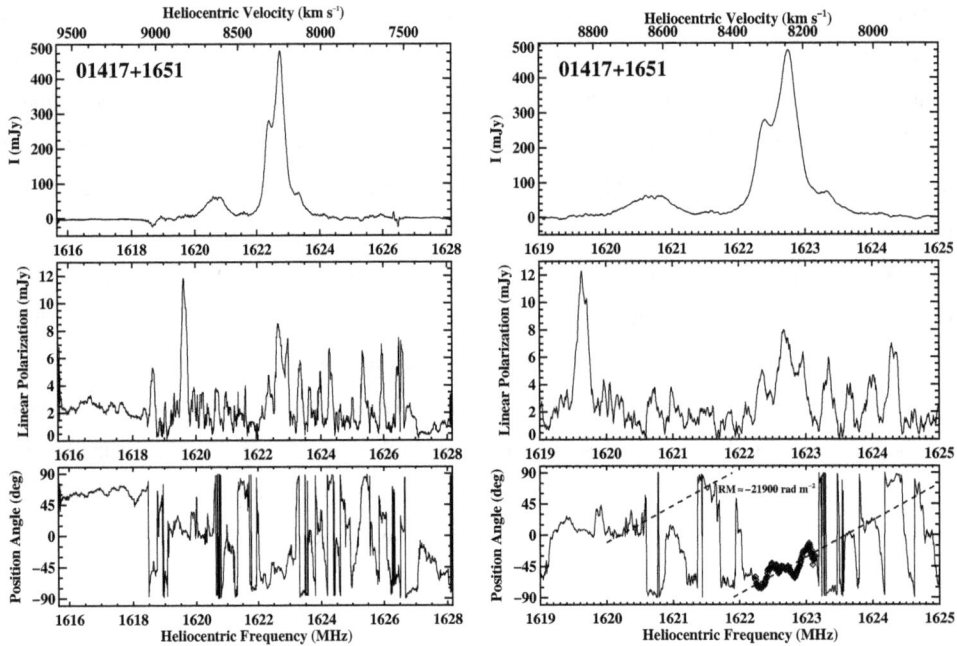

FIG. 4.9 —Linear polarization results for IRAS F01417+1651 (III Zw 35). *Top left*: Stokes *I*; *middle left*: linearly polarized intensity; *bottom left*: position angle over the entire 12.5 MHz bandwidth. *Right*: Same as left panels, but with the frequency range narrowed to 6 MHz; the bottom right panel also shows the fitted Faraday rotation as a dashed line whose slope was determined by fitting to the points marked as diamonds. All spectra are plotted as a function of heliocentric frequency (*bottom axis*). The top panels show the optical heliocentric velocity (*top axis*). All spectra are smoothed by a boxcar of 23 channels.

intensities, and the angle looks well defined for the 1622.8 MHz peak. Also, it seems to show a gradual change across the line, which is about 1 MHz wide. The dashed line displays the result of a least-squares fit to the frequency dependence of the position angle, using only those points that are marked as diamonds: RM $= -21,900 \pm 3700$ rad m^{-2}. The extrapolated dashed line goes through the clusters of points associated with spikes centered near 1623.7 and 1624.0, and moreover, even the slope of the line matches the data for these spikes. The slope also seems to match the 1624.4 MHz polarized-intensity spike, but the data are offset by about 60°. We speculate that (1) these three polarized-intensity spikes come from individual OHMs that are too weak to see in the top panels of Figure 4.9, (2) they all suffer the same Faraday rotation of $\simeq -21,900$ rad m^{-2} as the central peak, and (3) the intrinsic position angle for the 1624.4 MHz maser differs from the other two by about 60°.

4.5.2.2 IRAS F10173+0829

Figure 4.10 displays the linear polarization results for 10173. The polarized intensity shows a low-S/N spike that is centered on the Stokes I line: the polarization fraction is about 1% and the position angle about 60°. The spike is too narrow to fit for Faraday rotation. The hyperfine line ratio for 10173 is $R_H = 10.7$.

4.5.2.3 IRAS F12032+1707

The linear polarization results for 12032 are shown in Figure 4.11. The polarized intensity shows multiple spikes that might be real. The most significant is centered at $\simeq 1372$ MHz, with a peak flux density of $\simeq 5$ mJy, and has $\psi \simeq 15°$; near this frequency, I varies from $\simeq 10$ to $\simeq 40$ mJy, so if this peak is real, then the fractional polarization is huge, $\approx 50\%$ to $\approx 10\%$, unheard of for OH masers of any stripe.

4.5.2.4 IRAS F12112+0305

Figure 4.12 shows the linear polarization results for 12112. The lower frequencies are plagued by RFI, which remarkably disappears at the low-frequency boundary of the 1667 MHz line (centered at 1554.5 MHz). According to the National Telecommunications and Information Administration Manual of Regulations and Procedures for Federal Radio Frequency Management, this RFI is likely attributable to space-to-Earth aeronautical mobile satellite communications operated by Inmarsat. There is no trace of any detectable linear polarization for this source. This is the first

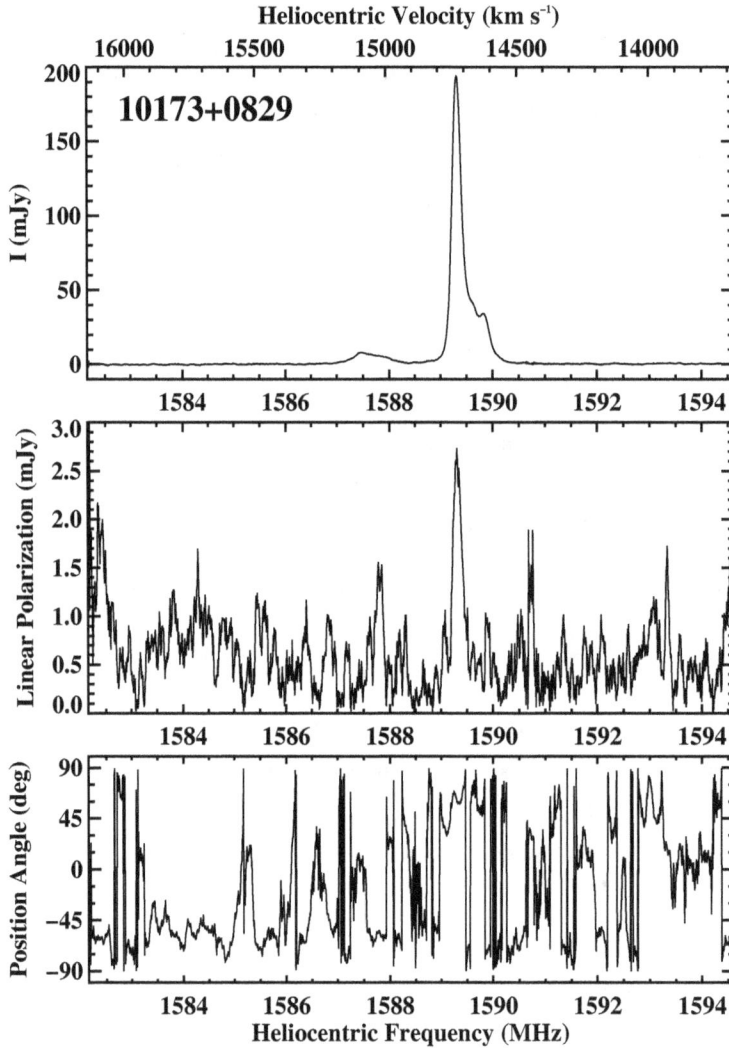

FIG. **4.10** — Linear polarization results for IRAS F10173+0829. See caption for Fig. 4.9. All spectra are smoothed by a boxcar of 17 channels.

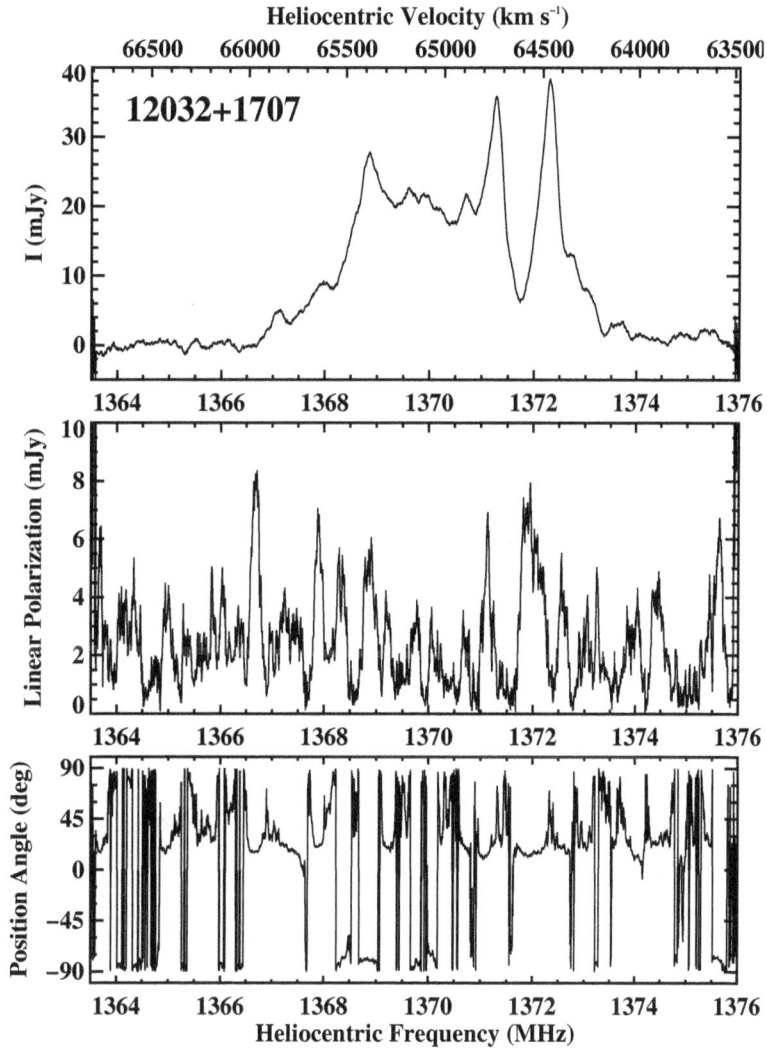

FIG. **4.11** — Linear polarization results for IRAS F12032+1707. See caption for Fig. 4.9. All spectra are smoothed by a boxcar of 31 channels.

detection of the 1665 MHz transition for 12112; the hyperfine line ratio is $R_H = 4.0$.

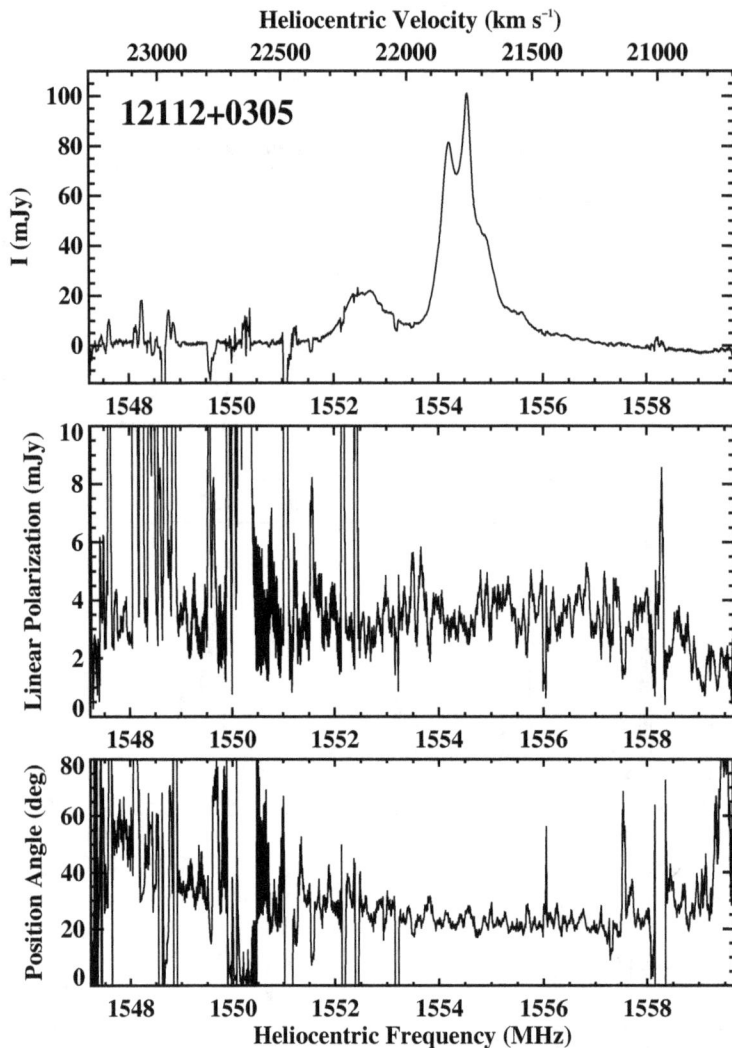

FIG. 4.12 — Linear polarization results for IRAS F12112+0305. See caption for Fig. 4.9. All spectra are smoothed by a boxcar of 11 channels.

4.5.2.5   IRAS F14070+0525

Figure 4.13 displays the linear polarization results for 14070. The linear polarization intensity is approximately 4 mJy across the entire 12.5 MHz bandwidth with an estimated position angle of $-48°$.

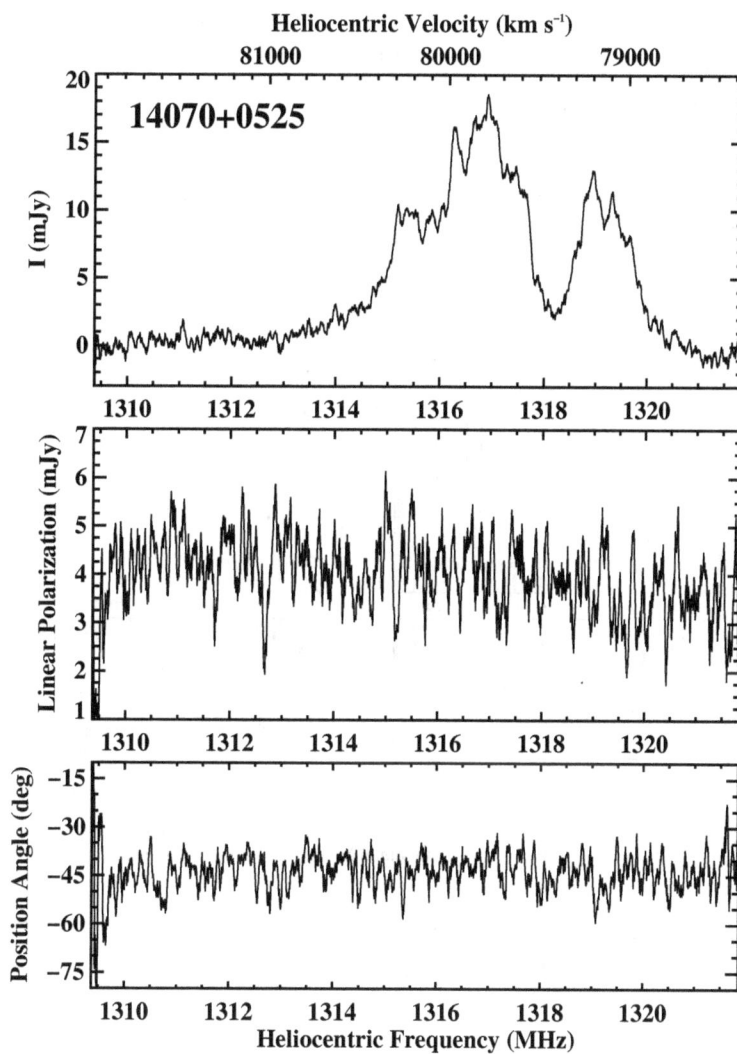

FIG. 4.13—Linear polarization results for IRAS F14070+0525. See caption for Fig. 4.9. All spectra are smoothed by a boxcar of 11 channels.

4.5.2.6   IRAS F15327+2340 (Arp 220)

The top panels of Figure 4.14 show the Stokes I profile for Arp 220 including both the 1665 and 1667 MHz transitions. The hyperfine line ratio is $R_H = 3.5$. The middle panels show the linear polarization intensity, which has a well-defined peak centered at 1638 MHz and peaks at about 2 mJy. This is only $\approx 0.3\%$ of the total intensity at this frequency. This is a very small fractional polarization but is very well detected.

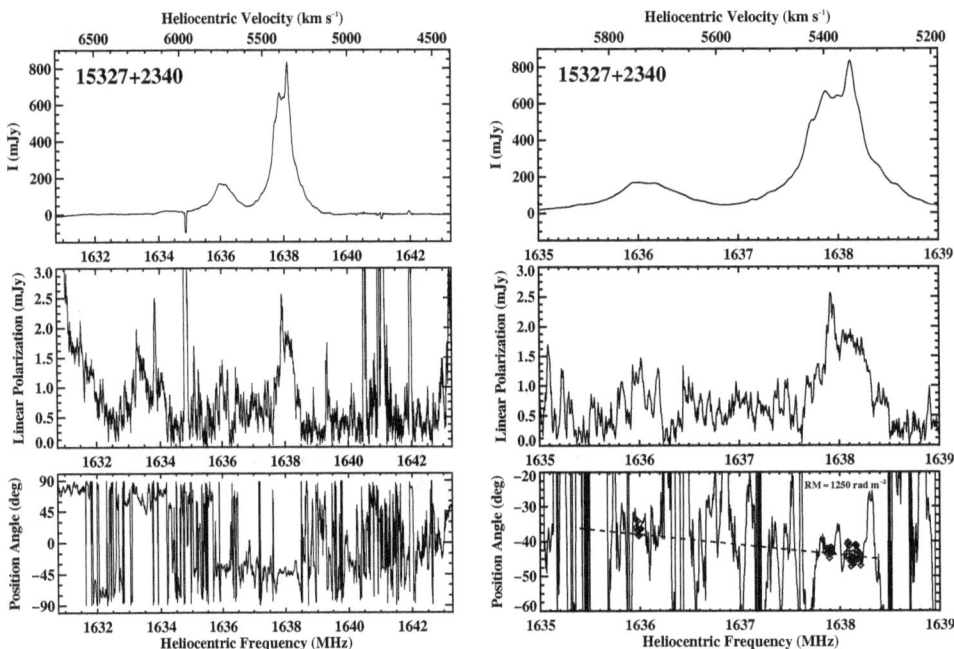

FIG. 4.14 — Linear polarization results for IRAS F15327+2340 (Arp 220). See caption for Fig. 4.9. *Right*: Frequency range has been narrowed to 4 MHz. All spectra are smoothed by a boxcar of nine channels.

The bottom panels show that the position angle of linear polarization is well defined in two regions of low noise, one centered near 1638 MHz and the other near 1636 MHz. The former region corresponds to the 1667 MHz line, and the latter is aligned with the 1665 MHz transition. This 1665 MHz line is unconvincingly visible in the polarized intensity spectrum, but the low noise in its position angle spectrum is unmistakable.

We fit the frequency variation of ψ to obtain the Faraday rotation measure RM using those points marked as diamonds in the bottom right panel of Figure 4.14. For the 1638 MHz component alone, we obtain RM $= 5230 \pm 7930$ rad m^{-2}. For the combination of the 1636 and 1638 MHz components, we obtain RM $= 1250 \pm 1040$ rad m^{-2}. These errors are considerable and make the formal result only marginally significant. The dashed line in the bottom right panel displays the result of the fit for both components together; visual inspection shows that not only is it an acceptable fit for both components together, but it is also acceptable for the 1638 MHz component alone. It is not unreasonable to conclude that the OHM radiation from both OH lines suffers a common Faraday rotation of RM $\approx$ 1250 rad m^{-2}; this is $\sim$20 times smaller than the value derived for III Zw 35.

4.6 Discussion

4.6.1 OH Maser Zeeman Pairs in the Milky Way

While our results are the very first in situ Zeeman detections in external galaxies, OH masers in the MW have been used as Zeeman magnetometers for well over a decade. In contrast to OHMs, Galactic OH maser emission lines are so narrow ($\sim$0.5 km s^{-1}) that fields of $\approx$1 mG are sufficient to completely split the left and right circular σ components into pairs. More than 100 of these Zeeman pairs have been compiled by Fish et al. (2003) and Reid & Silverstein (1990) with a distribution whose mean is consistent with 0 mG and whose standard deviation is 3.31$\pm$0.09 mG. Typical densities in OH maser regions are $n \sim 10^6$–10^7 cm^{-3}; for a field strength of $\sim$10 μG in gas at $\sim$1–100 cm^{-3}, the fields probed by Galactic OH masers are consistent with the enhancement of $|B| \propto n^{1/2}$ (Fish et al. 2003). The linear polarization of the σ components is often measured in addition to the π component, but the π components, which are in theory 100% linearly polarized, are rarely measured to be purely so.

Unlike in the OH masers in our Galaxy, the flux density of the 1667 MHz transition in all OHMs is larger than that of the 1665 MHz transition and, until now, no polarization has been detected (Lo 2005). There is no definitive explanation for the dominance of the 1667 MHz transition, but recent work suggests that this probably arises because the extragalactic lines are wider than the Galactic maser lines (P. Goldreich 2007, private communication; M. Elitzur 2007, private communication; Lockett & Elitzur 2008).

Our detections yield a median line-of-sight magnetic field strength of $\simeq$3 mG in OHMs in (U)LIRGs, which is comparable to the field strengths measured in OH masing regions in the MW. This strongly suggests that the *local* process of massive star formation occurs under similar conditions in (U)LIRGs, galaxies with vastly different large-scale environments than our own.

The magnetic field strengths we find in the OHMs in (U)LIRGs ($\sim$3 mG) are comparable to the volume-averaged fields of $\gtrsim$1 mG inferred from synchrotron observations. These results imply that milligauss magnetic fields likely pervade most phases of the interstellar medium (ISM) in (U)LIRGs. It is unclear, however, how to physically relate the two different magnetic field strengths in more detail given the possibility that each may probe rather different phases of the ISM. Some models of OHMs invoke radiative pumping in molecular clouds with gas densities $\sim 10^{3.5}$–10^4 cm^{-3} (e.g., Randell et al. 1995). This is similar to the *mean* gas density in the central ~ 100 pc in (U)LIRGs, in which case our observations likely probe the mean ISM magnetic field (whether the synchroton radiation also arises from gas at this density is unclear; upcoming *GLAST* observations of neutral pion decay may help assess this; see Thompson et al. 2007). It is also possible, however, that the OHMs arise in somewhat denser gas ($n \sim 10^6$–10^7 cm^{-3}; e.g., Lonsdale et al. 1998), as appears to be true in the MW (e.g., Fish et al. 2003). In this case, the magnetic field probed by OHMs is likely stronger than that in the bulk of the ISM. If we assume the $B \propto n^{1/2}$ scaling often assumed in the MW (Mouschovias 1976; Fish et al. 2003), the field strengths in the masing regions in (U)LIRGs are probably within a factor of ~ 3 of the mean ISM field (rather than a factor of several hundred in the MW), given the large mean gas densities in (U)LIRGs. This is still reasonably consistent with the mean field strength of $\gtrsim 1$ mG inferred from synchrotron observations. Without a better understanding of the physical conditions in the masing regions, however, it is difficult to provide a more quantitative connection between our inferred field strengths and either the mean ISM field or the magnetic field probed by synchrotron emission. Ultimately doing so is important because it will allow stringent constraints to be placed on the dynamical importance of magnetic fields across a wide range of physical conditions in (U)LIRGs.

Fish et al. (2003), with their comprehensive survey of Galactic OH masers and the accompanying statistical discussion, strongly support several previous suggestions that the field *direction* in OH masers usually mirrors that of the large-scale field in the vicinity of the masers. MERLIN observations of OH masers in Cep A by Bartkiewicz et al. (2005) also present Zeeman detections corroborating the field's alignment with the ambient ISM field direction. Thus, measuring the direction of the field in an OH maser reveals the field direction not only *in* the maser but also *outside and in the vicinity of* the OH maser. For the MW, this aids us to infer the large-scale magnetic field morphology. To directly compare the star formation processes in the MW and (U)LIRGs, it will be necessary to increase the sample of magnetic field strengths in (U)LIRGs and to directly map the Zeeman splitting of individual OHM spots using VLBI in order to probe whether reversals occur at smaller angular scales.

4.6.2 Strong Fields and Time Variability

Slysh & Migenes (2006) and Fish & Reid (2007) both observed fields of 40 mG using Zeeman observations of OH maser spots in W75N; these are the highest field strengths measured in Galactic OH masers and are an order of magnitude larger than the typical OH maser field. These OH maser spots also happen to have been flaring based on multiepoch VLBA observations; perhaps time variability in OH masers is correlated with strong magnetic fields. Interestingly, our strongest detection, $B_\parallel \sim 18$ mG in the gigamaser 12032, occurs in an OHM component that has increased in flux density by a factor of 2 since its previous published observation (Darling & Giovanelli 2001).

These results strongly support the development of an observational program to monitor both the time variability of the Stokes I flux density and magnetic field strength in OHMs as well as the necessity of observing the circular polarization of time-variable Galactic masers in hopes of detecting strong magnetic fields.

4.6.3 Linear Polarization and Faraday Rotation

Our measured rotation measures of RM $\simeq 21,900$ rad m^{-2} for III Zw 35 and $RM \simeq 1250$ rad m^{-2} for Arp 220 are large by most standards but are not unreasonable for (U)LIRGs. As mentioned in § 4.1, the magnetic field strength throughout the ULIRG ISM should be $\gtrsim 1$ mG from synchrotron observations. Electron densities are estimated to be ~ 1–10 cm^{-3} in the hot ionized plasma, both from observations of X-ray emission (e.g., Grimes et al. 2005) and from theoretical models of supernova-driven galactic winds (e.g., Chevalier & Clegg 1985). Over a path length of ~ 100 pc in the central portions of ULIRGs, $n_e \sim 1$ cm^{-3} and $B \sim 1$ mG imply $\langle n_e B_\parallel L \rangle \sim 0.1$ G cm^{-3} pc, or RM $\sim 80,000$ rad m^{-2}. This is a factor of 4–60 larger than our measured values.

It is reasonable for this simple estimate to overestimate the measured RM. This is because the RM depends only on the line-of-sight field component. The probability density function for the line-of-sight component of a randomly oriented magnetic field is flat between zero and the perfectly oriented case; thus, for a set of sources with randomly oriented fields, the observed line-of-sight field component is reduced by a factor of 2, and $\frac{1}{4}$ of the sources have the observed component less than $\frac{1}{4}$ the perfectly aligned value. In addition, and probably more importantly, the observed Faraday rotation responds only to the systematic line-of-sight field component, while the synchrotron radiation and the total magnetic energy depend on the total field, systematic plus random. Our estimate of RM $\sim 80,000$ is for the total field, not the systematic field, because the latter is much harder to predict.

The measured RM might also be reduced by finite source-size effects and/or propagation through an inhomogenous medium (Burn 1966). First, suppose that the magnetic field is everywhere uniform but that the Faraday rotation is produced in the same region where the maser radiation is produced, and that this region is extended along the line of sight. In this case, different line-of-sight depths of the maser are rotated by different amounts. This washes out the linear polarization and can reduce the apparent Faraday rotation. In the other extreme, think of the field as primarily random except for a small uniform component. Maser radiation observed at a given frequency might come from more than one maser located at different positions on the sky or at different distances into the source. In the former case, the RM might change with position on the sky; in the latter, it might change along the line of sight. In either case, its average value can be small. In addition, for an individual maser the field might fluctuate along the line of sight, reducing the total RM.

The interpretation of the linear polarization and RM data is thus currently difficult and nonunique. Observations of more systems would be helpful and may ultimately provide unique constraints on the thermal electron density and/or magnetic field structure (e.g., reversals) in the nuclei of (U)LIRGs.

Acknowledgments

It is a pleasure to acknowledge Phil Perillat, who performed the Mueller matrix calibration observations and reductions, measured the antenna gain, and wrote the online data acquisition software at Arecibo. We thank Karen O'Neil, Amy Shelton, and Mark Clark for helping us institute LSFS observing at the GBT. This research benefited from helpful discussions with Jeremy Darling, Vincent Fish, Peter Goldreich, Bill Watson, Fred Lo, Moshe Elitzur, Willem Baan, and Loris Magnani. T. R. appreciates the technical guidance of grammarian Elena Cotto. This research was supported in part by NSF grant AST-0406987. Support for this work was also provided by the NSF to T. R. through awards GSSP 05-0001, 05-0004, and 06-0003 from the NRAO. E. Q. was supported in part by NASA grant NNG06GI68G and the David and Lucile Packard Foundation. This research has made use of NASA's Astrophysics Data System Abstract Service and the SIMBAD database, operated at CDS, Strasbourg, France.

Chapter 5

Direct Detection of a Magnetic Field in a Galaxy at z = 0.692

Something unknown is doing we don't know what.

SIR ARTHUR STANLEY EDDINGTON, 1927

Content from this chapter has been submitted as a Letter to *Nature* with the following author list: Arthur M. Wolfe, Regina Jorgenson, Timothy Robishaw, Carl Heiles, & Jason X. Prochaska.

Abstract

We report the detection of Zeeman splitting in the absorption spectrum of a dampled Lyα system at $z = 0.692$ in the direction of 3C 286. This is the most distant in situ magnetic field ever measured in the universe. We find that the B field projected along the line of sight and averaged over a transverse dimension exceeding 200 pc is $\langle B_\parallel \rangle = 83.9 \pm 8.8$ μG. Our detection is completely unexpected because: (1) at a look-back time of 6.4 Gyr, this mean B field is 20 times larger than the current mean field; (2) this B field is embedded in neutral gas with a velocity dispersion of only 3.75 km s^{-1}; and (3) there is no evidence for the high star formation rates that normally accompany such high B fields in nearby galaxies. We present metallicity measurements for the system and consider the possibility that the field strength has been amplified by a shock.

5.1   Introduction

The magnetic field that permeates our Galaxy is a crucial constituent of the interstellar medium (ISM; Beck 2005). While B field strengths have been determined from measurements of Zeeman splitting of 21 cm absorption lines in interstellar H I clouds (Heiles & Troland 2004), and from measurements of Faraday rotation toward pulsars (Han et al. 2006), with one exception (Kazes et al. 1991; Sarma et al. 2005), such measurements have not been carried out for external galaxies. In those cases, equipartition between magnetic and cosmic-ray energy densities is assumed to infer B field strengths from their synchrotron emission. Although the resultant equipartition fields are comparable to the $\sim$5 μG average for mean field strengths in the Galaxy (Beck 2005), the assumption of equipartition remains untested and the origin of interstellar magnetic fields for any contemporary galaxy is still not understood. The leading theory is the mean-field dynamo model (Parker 1970) according to which large-scale mean magnetic fields in the past should be significantly weaker than $\sim$5 μG. In § 5.2, we discuss the first direct measurement of a magnetic field in a galaxy with a significant redshift, $z = 0.692$. We present our radio measurements in § 5.2 and optical metallicity measurements in § 5.3. In § 5.4 we discuss our results, and we provide a summary in § 5.5.

5.2   Radio Observations and Results

We detected Zeeman splitting of the 21 cm absorption line at $z = 0.6921534$ toward the quasar 3C 286 (Brown & Roberts 1973; Davis & May 1978). The absorption arises in a damped Lyα system (hereafter DLA-3C286), which is drawn from a population of neutral gas layers widely thought to be the progenitors of modern galaxies (Wolfe et al. 2005). The radio data for DLA-3C286 are summarized in Figure 5.1, which shows the line-depth spectra constructed from the $I(\nu)$ and $V(\nu)$ Stokes parameters near the 839.4 MHz frequency centroid of the redshifted 21 cm absorption line. The data were acquired in 12.6 hr of on-source integration with the 100 m Robert C. Byrd Green Bank Telescope[1] (GBT). Because the GBT feeds detect only orthogonal linearly polarized signals, while Zeeman splitting requires measuring circular polarization to construct Stokes $V(\nu)$, we generated $V(\nu)$ by cross-correlation techniques (Heiles et al. 2001b). The top panel of Figure 5.1 shows the $I(\nu)$ spectrum. A Gaussian fit to the absorption line yields a redshift

---

[1]The National Radio Astronomy Observatory is a facility of the National Science Foundation operated under cooperative agreement by Associated Observatories, Inc.

TABLE 5.1

PHYSICAL PARAMETERS FOR DLA-3C286 AND DLA-0235 INFERRED FROM 21 CM ABSORPTION

DLA	z	τ	σ_v[a] (km s^{-1})	$B_{\parallel}$[b] (μG)
3C 286	0.6921526 ± 0.0000008	0.095 ± 0.006	3.75 ± 0.20	83.9 ± 8.8
0235(a)	0.5239614 ± 0.0000003	0.428 ± 0.014	2.04 ± 0.06	6.8 ± 13.3
0235(b)	0.5239226 ± 0.0000004	0.538 ± 0.008	3.62 ± 0.14	15.3 ± 15.6
0235(c)	0.5238711 ± 0.0000005	0.342 ± 0.007	2.87 ± 0.09	16.1 ± 18.8
0235(d)	0.5237480 ± 0.0000002	0.421 ± 0.008	2.18 ± 0.04	0.12 ± 13.0

[a]Velocity dispersion of Gaussian fit to 21 cm absorption line.
[b]Line-of-sight component of B field inferred from Zeeman splitting.

z, central optical depth τ, and velocity dispersion[2] σ_v, all listed in Table 5.1, that are in good agreement with previous results (Brown & Roberts 1973; Davis & May 1978).

The signature of Zeeman splitting is searched for in the Stokes $V(\nu)$ spectrum, which equals the difference between the power in right-handed and left-handed circularly polarized signals, i.e., $V(\nu) = \text{RCP} - \text{LCP}$. The bottom panel of Figure 5.1 plots the line-depth function, $D_V(\nu) \equiv V(\nu)/I_c(\nu)$. In this case $D_V(\nu) = -[\tau_V(\nu)/2]\exp[-\tau(\nu)]$, where the difference between the optical depths of the RCP and LCP photons is defined as $\tau_V(\nu) \equiv \tau_{\text{RCP}}(\nu) - \tau_{\text{LCP}}(\nu) \ll 1$ (Heiles & Troland 2004). The $D_V(\nu)$ profile in Figure 5.1 shows the classic 'S curve' pattern expected for Zeeman splitting. This pattern is caused by the frequency offset between the $\tau_{\text{RCP}}(\nu)$ and $\tau_{\text{LCP}}(\nu)$ profiles, each being shifted from the unsplit line center by $\Delta\nu_z(z) = \pm 1.4(B_{\parallel}/\mu\text{G})(1+z)^{-1}$ Hz, respectively, where $B_{\parallel}$ is the B-field component projected along the line of sight. From our least-squares fit to the $D_V(\nu)$ pattern we find $B_{\parallel} = 83.9 \pm 8.8\ \mu$G (note, the direction of $B_{\parallel}$ is unknown since the instrumental sense of circular polarization was not calibrated). This magnetic field differs from the magnetic fields obtained from Zeeman splitting arising in interstellar clouds in the Galaxy in two respects. First, the field strength corresponds to the line-of-sight component of the mean field $\langle B_{\parallel} \rangle$ averaged over transverse dimensions exceeding 200 pc, since VLBI observations of the 21 cm absorption line show that the gas must extend more than $0''03$ across the sky in order to explain the difference between the velocity centroids of the fringe amplitude and phase-shift spectra (Wolfe et al. 1976). By contrast, the transverse dimensions of radio beams subtended at typical interstellar clouds is typically less than 1 pc. Second, this field strength is more than an order of magnitude stronger than the 6 μG average of B fields inferred from Zeeman splitting for

---

[2]The velocity dispersion is calculated using the optical radial velocity definition, $v_{\text{opt}}/c \equiv (\nu_0 - \nu)/\nu$, which differs from the radio definition, $v_{\text{rad}}/c \equiv (\nu_0 - \nu)/\nu_0$.

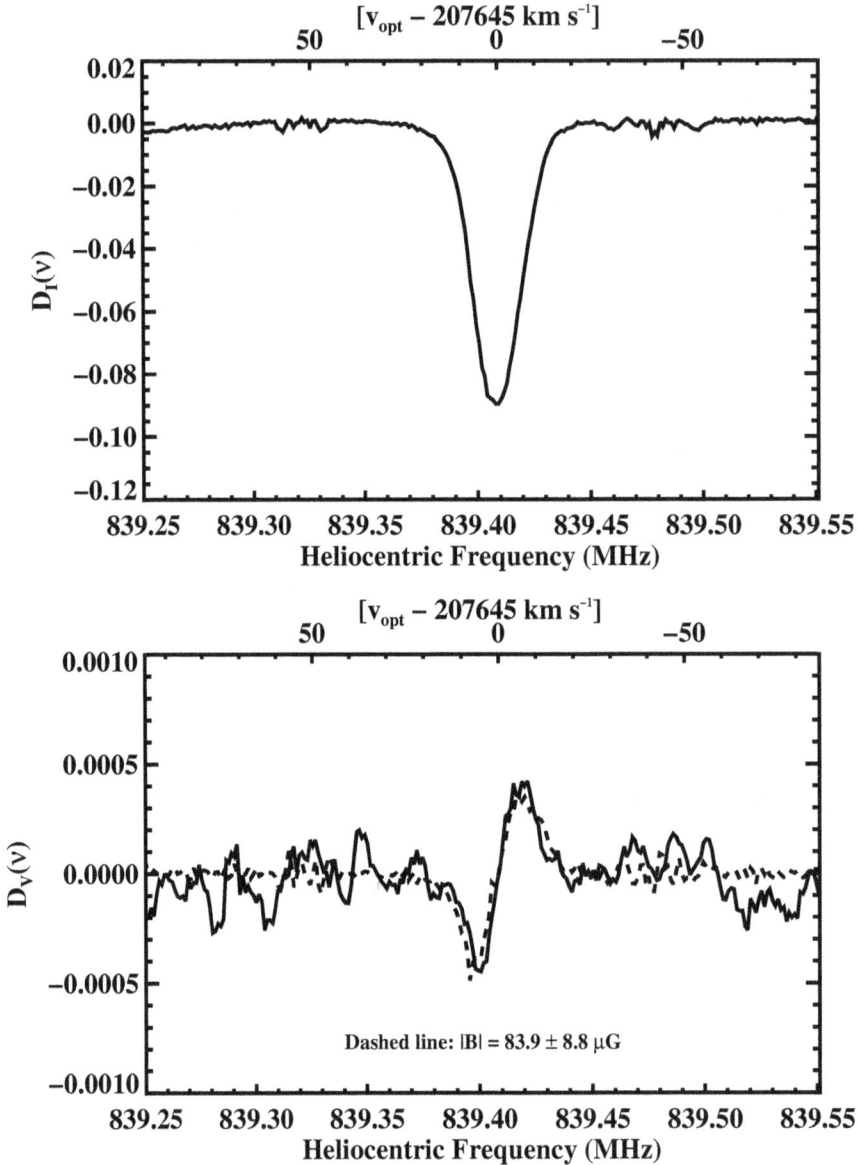

FIG. 5.1 — Stokes-parameter line-depth spectra for DLA-3C286. *Top*: Stokes $I(\nu)$, which equals the sum of orthogonal linear polarizations. Since $D_I(\nu) \equiv [I(\nu) - I_c(\nu)]/I_c(\nu)$, where $I_c(\nu)$ is a model fit to the $I(\nu)$ continuum, $D_I(\nu) = \exp[-\tau(\nu)] - 1$, where $\tau(\nu)$ is the average of optical depths in the two orthogonal states of linear polarization. *Bottom*: Stokes $V(\nu)$ over $I_c(\nu)$. Since Stokes $V(\nu)$ is the difference in the RCP and LCP intensities, the resulting $V(\nu)$ line profile is the difference between two Gaussian absorption profiles with frequency centroids shifted by $\pm\Delta\nu_z$ from the central frequency. The $V(\nu)$ profile resembles an 'S curve' because the RCP $-$ LCP intensity difference flips sign as it passes through the profile center. The dashed line shows the single-field fit for a Zeeman splitting caused by a line-of-sight magnetic field of strength 83.9$\pm$8.8 μG. The bottom axis of each plot is the heliocentric frequency, and the top axis shows the optical heliocentric velocity relative to that of the line center for DLA-3C286.

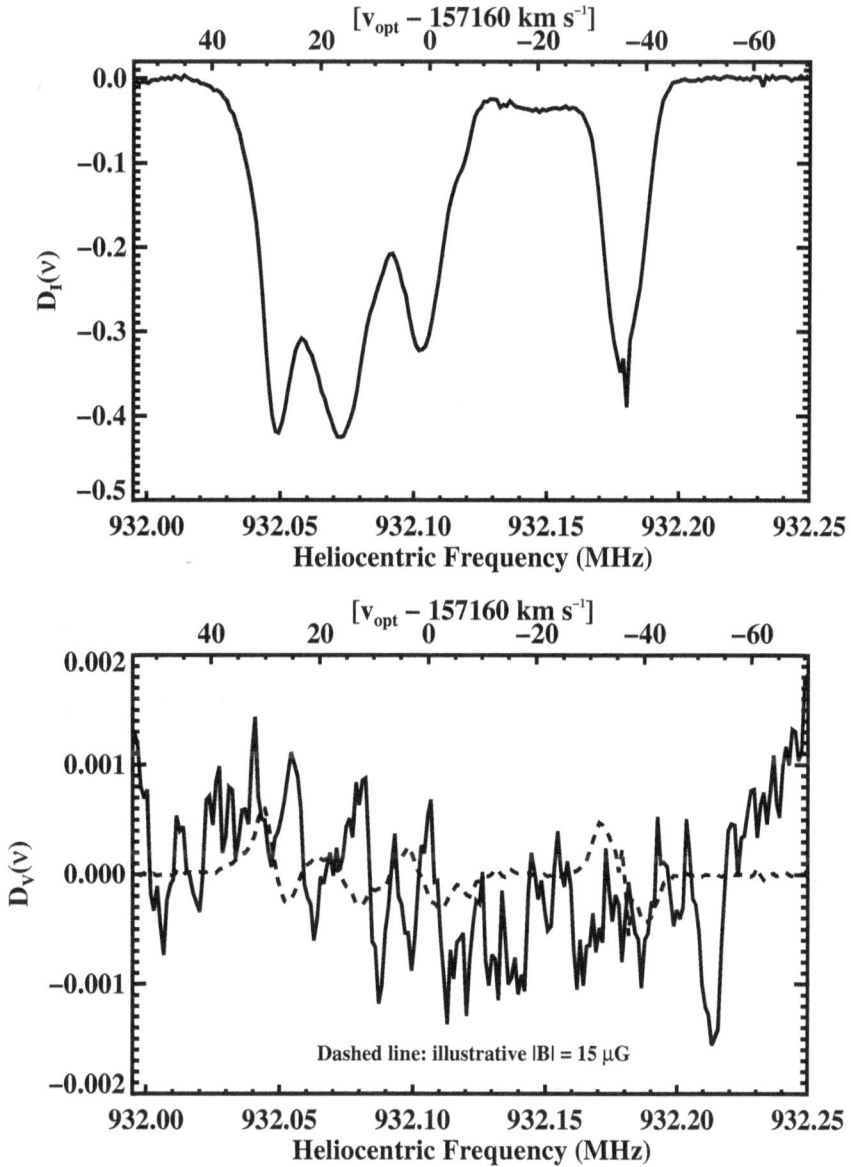

FIG. 5.2 — Stokes-parameter line-depth spectra for DLA-0235. See caption for Fig. 5.1. *Top*: There are four absorption features from this DLA, corresponding with increasing frequency to the components (a)–(d) in Table 5.1. *Bottom*: No Zeeman detection is evident in the Stokes V line-depth spectrum. The dashed line is the expected Stokes V profile for a single line-of-sight magnetic field of strength 15 μG. The top axis of each plot shows the optical heliocentric velocity relative to that of the line center for component (c) of DLA-0235.

TABLE 5.2

PHYSICAL PARAMETERS FOR DLA-3C286 INFERRED FROM OPTICAL ABSORPTION

Ion X	$\log_{10} N(X)$ (cm^{-2})	$[X/H]$
H I..........	21.25 ± 0.02	$\cdots$
Fe II	15.09 ± 0.01	-1.66 ± 0.02
Cr II	13.44 ± 0.01	-1.48 ± 0.02
Zn II	12.53 ± 0.03	-1.39 ± 0.03
Si II	>15.48	>-1.31

interstellar clouds (Heiles & Troland 2004), but is comparable to the total magnetic fields deduced from equipartition arguments for the nuclear rings in barred spiral galaxies (Beck 2005).

We also searched for Zeeman splitting in the Stokes V line-depth spectrum of the 21 cm absorption feature arising at $z = 0.524$ toward the Blazar AO 0235+16 (Roberts et al. 1976). This absorption line arises in a DLA (hereafter DLA-0235 Cohen et al. 1987) associated with a spectroscopically identified star-forming galaxy (Burbidge et al. 1996). The data were acquired in 9.76 hr of on-source integration time during the same observing run from which our DLA-3C286 results were obtained. However, the DLA-0235 data are of poorer quality because interference near the absorption frequencies was so strong that the useful integration time was significantly reduced. The results plotted in the top panel of Figure 5.2 show four velocity components in the $D_I(\nu)$ spectrum, but no S-curve patterns at the frequency centroids of these components in the $D_V(\nu)$ spectrum in the bottom panel. Our least-squares fits for Zeeman splitting in each of the components yielded statistically insignificant measurements for B_{los}. The derived parameters are shown in Table 5.1. The results imply 95% confidence upper limits as low as ≈ 25 μG on $B_\parallel$ for some of the components in this DLA.

5.3 Optical Observations and Results

We obtained further information about conditions in the absorbing gas in DLA-3C286 from accurate optical spectra acquired with the HIRES Echelle spectrograph on the Keck I 10 m telescope.[3] Figure 5.3 shows velocity profiles for several resonance absorption lines arising from dominant low ions associated with the 21 cm absorption system. The results of our least-squares fit to the velocity profiles with a Gaussian velocity distribution are shown in Table 5.2, where the

[3]The W. M. Keck Observatory is operated as a scientific partnership among the California Institute of Technology, the University of California, and the National Aeronautics and Space Administration. The Observatory was made possible by the generous financial support of the W. M. Keck Foundation.

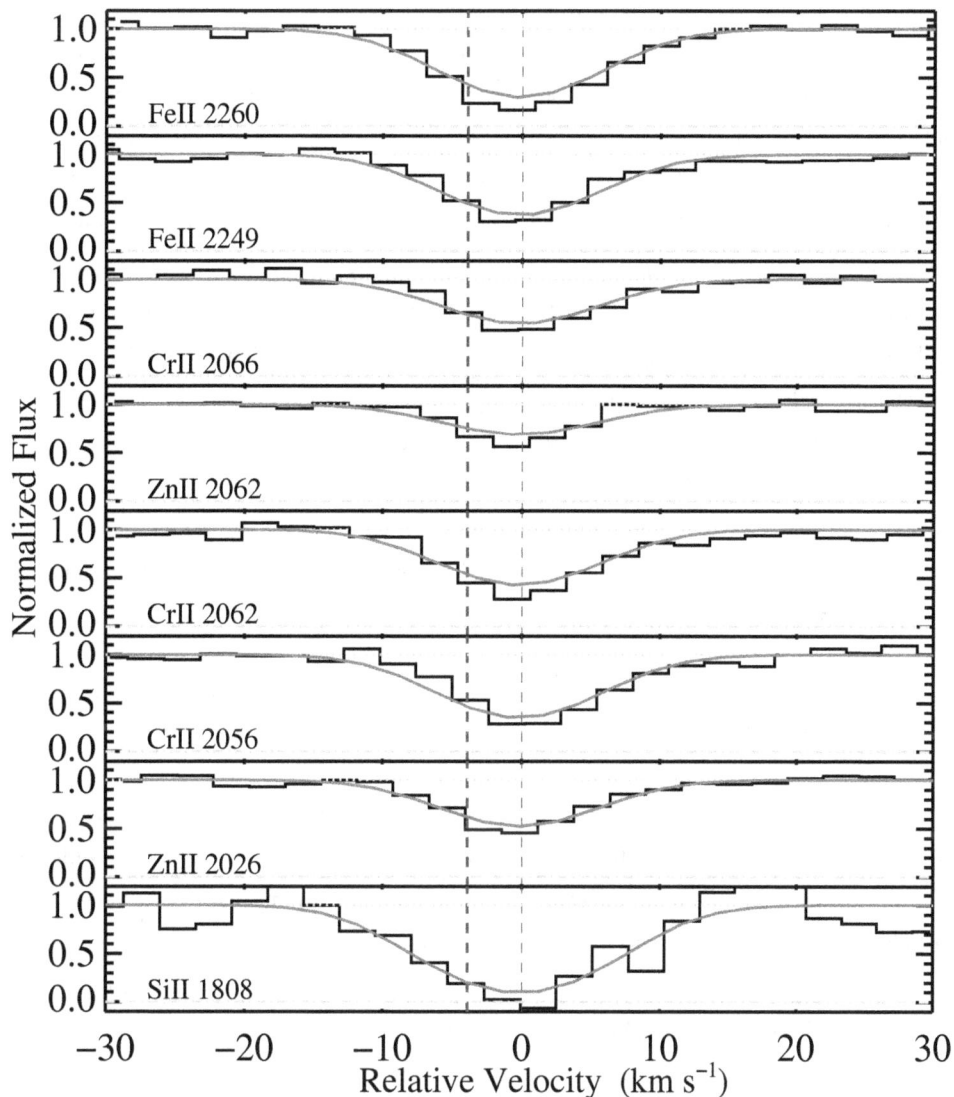

FIG. 5.3 — HIRES velocity profiles for dominant low ions in the 21 cm absorber toward 3C 286. Spectral resolution is $\Delta v = 7.0$ km s^{-1} and the average signal-to-noise ratio per 2.1 km s^{-1} pixel is about 30:1. The bold dashed vertical line at -3.8 km s^{-1} denotes the velocity centroid of the single-dish 21 cm absorption feature, and the faint dashed vertical line denotes the velocity centroid of the resonance lines shown for each ion. The histograms show the measured data and the smooth lines the Gaussian fits to each profile.

optical redshift is displaced $+3.8$ km s^{-1} from the 21 cm redshift in Table 5.1, and σ_v is significantly narrower than σ_v obtained from the 21 cm profile (this is likely due to the ≈ 1 pc transverse
size of the optical beam to the quasar, which penetrates a subset of clouds encountered by the 200
pc transverse size of the radio beam [see § 5.4]). This solution also yields ionic column densities
from which we derived the metal abundances in Table 5.2. Because refractory elements such as Fe
and Cr can be depleted onto dust grains (Savage & Sembach 1996), we used the volatile elements
Si and Zn to derive a logarithmic metal abundance with respect to solar of [M/H] $= -1.30$. The
depletion ratios [Fe/Si] and [Cr/Zn] were then utilized to derive a conservative upper limit on the
logarithmic dust-to-gas ratio relative to Galactic values of [D/G] < -1.8. These are lower than
the mean values of [M/H] and [D/G] deduced for DLAs at $z = 0.7$ (Meiring et al. 2006; Wolfe
et al. 2003). The low metallicity indicates a history of low star formation rates (SFRs). Because
the intensity of FUV radiation emitted by young massive stars is proportional to the SFR, low
SFRs should result in low grain photoelectric heating rates per H atom, Γ_{pe} (Wolfe et al. 2003).
This is consistent with the low upper limit, $\Gamma_{pe} < 10^{-27.4}$ erg s^{-1} H^{-1}, obtained by combining
the assumption of thermal balance with the absence of C II* $\lambda 1335.7$ absorption in the previous
low-resolution HST spectra of 3C 286 (Boisse et al. 1998), and indicates an SFR per unit area
Σ_{SFR} less than the solar neighborhood value of 0.004 M$_\odot$ yr^{-1} kpc^{-2} (Kennicutt 1998).

5.4   Discussion

We have detected an unusually strong magnetic field at $z = 0.692$ with a coherence length
exceeding 200 pc in neutral gas that is quiescent, metal-poor, nearly dust-free, and with little,
if any, evidence of star formation. Furthermore, the gas is highly magnetized, as suggested by
the value of the 'notional' mass-to-magnetic-flux ratio λ_n, which indicates that the magnetic field
alone prevents gravitational collapse. Assuming a plausible configuration in which the gas is in a
plane parallel sheet with half-height H, in-plane magnetic field B_{plane}, and mass surface density Σ,
one finds that $\lambda_n = 2\pi G^{1/2}/B_{plane}$ (Shu et al. 2007). Assuming a disk inclination angle $i = 45°$,
we have $\Sigma = 13$ M$_\odot$ pc^{-2} and $B_{plane} = 119$ G, in which case $\lambda_n = 0.038$ for a mean molecular
weight $\mu = 1.3$. Because the physical mass-to-magnetic-flux ratio, λ, depends on B_{norm}, the
magnetic field parallel to the sheet normal, rather than B_{plane}, one finds that $\lambda = (L/H)\lambda_n$, where
L is the length scale in the plane for gas with a fixed volume density (Shu et al. 2007). As a result,
L must exceed ≈ 5 kpc for $\lambda > 1$ and for collapse to occur, which is unrealistically large for the
sizes of self-gravitating configurations in a galaxy. Therefore the subcritical gas in DLA-3C286 is

even more strongly magnetized than the cold neutral medium (CNM) gas in the Galaxy where the mean $\langle \lambda_n \rangle < 0.16$ (McKee & Ostriker 2007).

To intepet the type of object we have detected, we first focus on the low value of σ_v (=3.8 km s^{-1}). This brings to mind the outskirts of face-on spirals, where 21 cm emission profiles show that σ_v decreases to a universal value of 6 km s^{-1} beyond the deVaucouleurs radii of these galaxies (Dickey et al. 1990). Indeed, the outer regions of the low surface-brightness galaxy detected $2''\!.5$ (17.8 kpc at $z = 0.692$) from 3C 286 may be associated with the 21 cm absorber (Steidel et al. 1994). Since such galaxies exhibit sub-solar metallicities, the low value of [M/H] in DLA-3C286 strengthens the case for this association. But such a configuration does not satisfy magnetostatic equilibrium, in which the total midplane pressure, $B_{\mathrm{plane}}^2/8\pi + \sigma_v^2 \rho$, equals the 'weight' of the gas, $\pi G \Sigma^2/2$, because the pressure-to-weight ratio, which reaches a minimum for the assumed inclination angle $i = 45°$, equals 715. Therefore, self-consistent magnetostatic configurations are ruled out unless the stellar contribution to Σ exceeds ≈ 350 M$_\odot$ pc^{-2}. This is larger than the 50 M$_\odot$ pc^{-2} surface density perpendicular to the solar neighborhood, and is a factor of 100 or more higher than the surface densities of stars detected in the outer regions of low surface-brightness galaxies (de Blok et al. 1995). Although high stellar surface densities are present in the central regions of galaxies, and probably confine the highly magnetized gas in the nuclear rings of barred spirals exhibiting *total* field strengths inferred from equipartition arguments to be ~100 μG (Beck 2005), the rings are associated with regions of active star formation and high molecular and dust content. Consequently, these are unlikely sites of the B field detected in DLA-3C286. On the other hand, the absorption site might be highly magnetized gas confined by the gravity exerted by a disk of old stars; i.e., a disk of stars and gas in another galaxy located toward the bright background quasar. The H I disks found at the centers of elliptical galaxies (Morganti et al. 2006) or the central regions of disks in early-type spiral galaxies are possible prototypes. Perhaps the strong B field was generated during a star-forming phase at $z \gg 0.692$. However, the time interval of the star-forming event during which the large B field was generated is constrained by the low metallicity of the gas.

Because of the uncertainties, we consider the possibility that the B field observed in DLA-3C286 does not reflect average conditions in the associated galaxy, but rather is enhanced by a shock (F. H. Shu 2008, private communication). Let B_1 and B_2 be components of the B field parallel to the shock front in the pre-shock and post-shock gas. In the limit of flux freezing in a highly magnetized 1-D radiative shock, one finds that $u_1 = (2B_1/B_2)^{-1/2} v_{\mathrm{A2}}$ where u_1 is the velocity of the pre-shock gas relative to the shock front, and the Alfvén speed of the post-shock gas is given by $v_{A2} \equiv B_2(4\pi\rho_2)^{-1/2}$. If the shock moves perpendicular to the plane, then $B_2 = B_{\text{plane}} = 119\ \mu\text{G}$ and $v_{A2} \geq 72\ \text{km s}^{-1}$, since our two-phase model estimates indicate $n_2 \leq 10$ cm^{-3} (Wolfe et al. 2003). Assuming a typical value of $B_1 = 5\ \mu\text{G}$, we find $u_1 \geq 250\ \text{km s}^{-1}$, which is a reasonable limit for the impact velocity generated by a merger between the gaseous disks of two late-type galaxies. The possible detection of a second galaxy within $1''$ of 3C 286 in a WFPC-2 image (Le Brun et al. 1997) lends some support to this idea. But the problem with this scenario is that the second disk would create another set of absorption lines displaced ≥ 250 km s^{-1} from the redshift of DLA-3C286, which is the only redshift observed.

By contrast a merger with an elliptical galaxy could result in only one DLA redshift, since a significant fraction of ellipticals do not contain H I disks (Morganti et al. 2006). In this case a shock front moving in the plane of the first galaxy would be generated by the gravitational impulse induced by the elliptical moving normal to the plane. Assuming the mass distribution of the elliptical is described by a singular isothermal sphere with velocity dispersion σ_{int} (Koopmans et al. 2006), we find the gas responds to the impulse with an inward radial velocity (Binney & Tremaine 1987) $\Delta v_R = -2\pi\sigma_{\text{int}}^2/v_{\text{int}}$, where v_{int} is the velocity of the elliptical normal to the plane. Thus, an average elliptical galaxy with $\sigma_{\text{int}} = 80$ km s^{-1}, and mass $M = 2 \times 10^{11}$ M$_\odot$ at $z = 0.7$, would produce $\Delta v_R = -125$ km s^{-1} if $v_{\text{int}} = 300$ km s^{-1} . This would give rise to an outwardly propagating cylindrical shock front with speed $u_1 \approx 2|\Delta v_R| \approx 250$ km s^{-1}. Since B field enhancement only occurs in directions in which components of B_{plane} are parallel to the shock front, magnetic-field enhancement would not be as strong as when the shock moves perpendicular to the plane. Therefore, B_{plane} in the pre-shock gas would have to exceed 5 μG to explain the strength of the observed Zeeman splitting. Since λ_n is an invariant for fluid flow across shock fronts, this unshocked average gas must also be highly magnetized. Therefore, low mass-to-magnetic-flux ratios may be generic features of low surface-brightness galaxies, and may be the reason behind their low SFRs (de Blok et al. 1995). While this is a promising scenario, one still must explain why the post-shock velocity field averaged over length scales of 200 pc produces narrow absorption lines with $\sigma_v = 3.75$ km s^{-1}. This is a crucial constraint for any model.

5.5 Summary

We have made the first direct measurement of a magnetic field at a significant redshift. We find that B is stronger than 83.9 μG in a galaxy at $z = 0.692$. This result is completely unexpected, because dynamo theory predicts that large-scale mean fields in galaxies should be weaker rather than stronger in the past (Parker 1970), and since mean B fields this strong have normally been inferred for the star-forming regions near the centers of galaxies, rather than for the quiescent, metal-poor gas with low SFRs that we have detected. Because the B field in DLA-3C286 cannot be confined by the self-gravity of the gas, we considered the possibility that the gas is confined by the gravity exerted by an older population of disk stars, or that the B field is enhanced by shocks. The shocks might be generated by the gravitational impulse set off by a merger between a low surface-brightness galaxy with ambient B field exceeding ~ 10 μG and a gas-free elliptical galaxy. One benefit of this scenario is that the mean pre-collision B field in DLA-3C286 would only be required to build up to 10 μG rather than 100 μG. But it is unclear whether the dynamo mechanism acting on plausible seed fields can build up to such field strengths in the 4 to 5 Gyr age of this DLA. While the similarity between the FIR-radio correlations at $z \sim 1$ and $z = 0$ (Appleton et al. 2004) is tentative evidence that ~ 10 μG fields are present at $z = 0.692$, these are fields derived by assuming equipartition. However, our first direct measurement of a B field at $z \gg 0$ indicates the magnetic energy density greatly exceeds the turbulent kinetic and presumably cosmic-ray energy densities.

Finally, in the case of the merger scenario, the probability of detecting ~ 100 μG fields in a random sample of 21 cm absorbers is $p = R_{\mathrm{merge}}(z)t_{\mathrm{dur}}$, where $R_{\mathrm{merge}}(z)$ is the merger rate per galaxy at redshift z, and t_{dur} is the timescale for the duration of B field enhancement. This is given by the impulse duration time, $t_{\mathrm{impulse}} \approx 2R/v_{\mathrm{int}}$, plus the timescale for a rarefaction wave to propagate a distance R from the impact site to the quasar radio beam, i.e., $t_{\mathrm{rare}} \approx R/v_{\mathrm{A2}}$. If $R \approx 3$ kpc, we have $t_{\mathrm{dur}} \approx 6.5 \times 10^7$ yr. Since the the fraction of major mergers $f_m = 10 \pm 2\%$ at $0.2 < z < 1.2$, and assuming a merger timescale $t_m = 0.5$–1 Gyr, we find $R_{\mathrm{merge}} = (1$–$2) \times 10^{-10}$ yr^{-1} (Lotz et al. 2008). Given the uncertainties in t_{dur}, we estimate $p = 0.006$–0.03; i.e., the probability for detecting such events is low. This is consistent with our null detection of a strong B field in another 21 cm absorber, DLA-0235, that we observed along with DLA-3C286 (see Figure 5.2 and Table 5.1). Of course p might be enhanced if the sample were limited to 21 cm absorbers selected for signatures of high B fields; e.g., 21 cm line widths as narrow as in DLA-3C286.

Acknowledgments

We wish to thank Frank H. Shu for suggesting the merger model. We also thank Frank H. Shu, Eric Gawiser, and Alex Lazarian for valuable comments. The Keck observers wish to extend special thanks to those of Hawaiian ancestry on whose sacred mountain we are privileged to be guests. Without their generous hospitality, none of the optical observations presented here would have been possible. A. M. W. and J. X. P. were partially supported by NSF grant AST 07-09235. C. H. and T. R. were partially supported by NSF grant AST-0406987. Support for this work was also provided by the NSF to T. R. through awards GSSP 05-001, 05-004, 06-003 from the NRAO.

One must stop *before* one has finished;
otherwise, one will never stop and never finish.

BARBARA W. TUCHMAN, 1963

136

Bibliography

Of the 99.9% of the ApJ articles you don't read, 99.8% of them are wrong. So why waste your time?

CARL HEILES, 2003

Appleton, P. N., et al. 2004, ApJS, 154, 147

Baan, W. A., Rhoads, J., Fisher, K., Altschuler, D. R., & Haschick, A. 1992, ApJ, 396, L99

Baan, W. A., Wood, P. A. D., & Haschick, A. D. 1982, ApJ, 260, L49

Babcock, H. W. 1947, PASP, 59, 112

Ball, J. A., & Meeks, M. L. 1968, ApJ, 153, 577

Barrett, A. H., & Rogers, A. E. E. 1966, Nature, 210, 188

Bartkiewicz, A., Szymczak, M., Cohen, R. J., & Richards, A. M. S. 2005, MNRAS, 361, 623

Beck, R. 2005, in Cosmic Magnetic Fields, ed. R. Wielebinski & R. Beck (Berlin: Springer), 41

Beck, R., & Gaensler, B. M. 2004, New Astronomy Review, 48, 1289

Binney, J., & Tremaine, S. 1987, Galactic Dynamics (Princeton: Princeton Univ. Press)

Boisse, P., Le Brun, V., Bergeron, J., & Deharveng, J.-M. 1998, A&A, 333, 841

Bolton, J. G., & Wild, J. P. 1957, ApJ, 125, 296

Born, M., & Wolf, E. 1999, Principles of Optics (7th ed.; Cambridge: Cambridge Univ. Press)

Bot, C., Boulanger, F., Rubio, M., & Rantakyro, F. 2007, A&A, 471, 103

Bourke, T. L., Myers, P. C., Robinson, G., & Hyland, A. R. 2001, ApJ, 554, 916

Bregman, J. D., Troland, T. H., Forster, J. R., Schwarz, U. J., Goss, W. M., & Heiles, C. 1983, A&A, 118, 157

Brooks, J. W., Murray, J. D., & Radhakrishnan, V. 1971, Astrophys. Lett., 8, 121

Brown, R. L., & Roberts, M. S. 1973, ApJ, 184, L7

Burbidge, E. M., Beaver, E. A., Cohen, R. D., Junkkarinen, V. T., & Lyons, R. W. 1996, AJ, 112, 2533

Burke, B. F., & Graham-Smith, F. 1997, An Introduction to Radio Astronomy (1st ed.; Cambridge: Cambridge Univ. Press)

Burn, B. J. 1966, MNRAS, 133, 67

Chandrasekhar, S. 1947, ApJ, 105, 424

————. 1989, in Selected Works, Vol. 2 (Chicago: Univ. Chicago Press), 511

Chandrasekhar, S., & Fermi, E. 1953, ApJ, 118, 116

Chauvenet, W. 1960, A Manual of Spherical and Practical Astronomy, Vol. 2 (5th ed.; New York: Dover)

Chevalier, R. A., & Clegg, A. W. 1985, Nature, 317, 44

Christiansen, W. N., & Högbom, J. A. 1969, Radiotelescopes (Cambridge: Cambridge Univ. Press)

Chu, T.-S., & Turrin, R. 1973, IEEE Trans. Antenn. Prop., 21, 339

Cohen, M. H. 1958, Proc. IRE, 46, 172

Cohen, R. D., Smith, H. E., Junkkarinen, V. T., & Burbidge, E. M. 1987, ApJ, 318, 577

Coles, W. A., & Rumsey, V. H. 1970, ApJ, 159, 247

Coles, W. A., Rumsey, V. H., & Welch, W. J. 1968, ApJ, 154, L61

Condon, E. U., & Shortley, G. H. 1957, The Theory of Atomic Spectra (Cambridge: Cambridge Univ. Press)

Condon, J. J., Huang, Z.-P., Yin, Q. F., & Thuan, T. X. 1991, ApJ, 378, 65

Conway, R. G. 1974, in Planets, Stars and Nebulae Studied with Photopolarimetry, ed. T. Gehrels (Tucson: Univ. Arizona Press), 352

Cox, A. N. 2000, Allen's Astrophysical Quantities (4th ed.; New York: AIP Press)

Crutcher, R. M. 1999, ApJ, 520, 706

Crutcher, R. M., Evans, N. J., II, Troland, T., & Heiles, C. 1975, ApJ, 198, 91

Crutcher, R. M., & Kazes, I. 1983, A&A, 125, L23

Crutcher, R. M., Nutter, D. J., Ward-Thompson, D., & Kirk, J. M. 2004, ApJ, 600, 279

Crutcher, R. M., & Troland, T. H. 2000, ApJ, 537, L139

Crutcher, R. M., Troland, T. H., Goodman, A. A., Heiles, C., Kazes, I., & Myers, P. C. 1993, ApJ, 407, 175

Crutcher, R. M., Troland, T. H., & Heiles, C. 1981, ApJ, 249, 134

Crutcher, R. M., Troland, T. H., & Kazes, I. 1987, A&A, 181, 119

Crutcher, R. M., Troland, T. H., Lazareff, B., & Kazes, I. 1996, ApJ, 456, 217

Crutcher, R. M., Troland, T. H., Lazareff, B., Paubert, G., & Kazès, I. 1999, ApJ, 514, L121

Dame, T. M., Hartmann, D., & Thaddeus, P. 2001, ApJ, 547, 792

Darling, J., & Giovanelli, R. 2000, AJ, 119, 3003

————. 2001, AJ, 121, 1278

————. 2002, AJ, 124, 100

Davies, R. D. 1974, in IAU Symp. 60, Galactic Radio Astronomy, ed. F. J. Kerr & S. C. Simonson (Dordrecht: Reidel), 275

———. 1994, in Cosmical Magnetism, ed. D. Lynden-Bell (Dordrecht: Kluwer), 131

Davies, R. D., Booth, R. S., & Wilson, A. J. 1968, Nature, 220, 1207

Davies, R. D., de Jager, U., & Verschuur, U. L. 1966, Nature, 209, 974

Davis, M. M., & May, L. S. 1978, ApJ, 219, 1

de Blok, W. J. G., van der Hulst, J. M., & Bothun, G. D. 1995, MNRAS, 274, 235

del Toro Iniesta, J. C. 2003, Introduction to Spectropolarimetry (Cambridge: Cambridge Univ. Press)

Diamond, P. J., Lonsdale, C. J., Lonsdale, C. J., & Smith, H. E. 1999, ApJ, 511, 178

Dickey, J. M., Hanson, M. M., & Helou, G. 1990, ApJ, 352, 522

Downes, D., & Solomon, P. M. 1998, ApJ, 507, 615

Enz, C. P. 2002, No Time To Be Brief - A Scientific Biography of Wolfgang Pauli (Oxford: Oxford Univ. Press)

Falgarone, E., Troland, T. H., Crutcher, R. M., & Paubert, G. 2008, A&A, submitted

Fiebig, D., & Güesten, R. 1989, A&A, 214, 333

Fish, V. L., & Reid, M. J. 2006, ApJS, 164, 99

———. 2007, ApJ, 656, 952

Fish, V. L., Reid, M. J., Argon, A. L., & Menten, K. M. 2003, ApJ, 596, 328

Forman, P. 1970, in Historical Studies in the Physical Sciences, Vol. 2, ed. R. McCormmach (Philadelphia: Univ. Pennsylvania Press), 153

Goldsmith, P. F., Heyer, M., Narayanan, G., Snell, R., Li, D., & Brunt, C. 2008, ApJ, in press (arXiv:0802.2206)

Goldstein, D. 2003, Polarized Light (New York: Marcel Dekker)

Goodman, A. A., Crutcher, R. M., Heiles, C., Myers, P. C., & Troland, T. H. 1989, ApJ, 338, L61

Goodman, A. A., & Heiles, C. 1994, ApJ, 424, 208

Graham, J. R., Carico, D. P., Matthews, K., Neugebauer, G., Soifer, B. T., & Wilson, T. D. 1990, ApJ, 354, L5

Green, R. M. 1985, Spherical Astronomy (Cambridge: Cambdridge Univ. Press)

Greve, A., & Pauls, T. 1980, A&A, 82, 388

Grimes, J. P., Heckman, T., Strickland, D., & Ptak, A. 2005, ApJ, 628, 187

Güesten, R., Fiebig, D., & Uchida, K. I. 1994, A&A, 286, L51

Güsten, R., & Fiebig, D. 1990, in IAU Symp. 140, Galactic and Intergalactic Magnetic Fields, ed. R. Beck, R. Wielebinski, & P. P. Kronberg (Dordrecht: Kluwer), 305

Hale, G. E. 1908, ApJ, 28, 315

Hamaker, J. P., & Bregman, J. D. 1996, A&AS, 117, 161

Han, J. L., Manchester, R. N., Lyne, A. G., Qiao, G. J., & van Straten, W. 2006, ApJ, 642, 868

Hartmann, D., & Burton, W. B. 1997, Atlas of Galactic Neutral Hydrogen (Cambridge: Cambdridge Univ. Press)

Heiles, C. 1987, in Interstellar Processes, ed. D. J. Hollenbach & H. A. Thronson Jr. (Dordrecht: Reidel), 171

———. 1988, ApJ, 324, 321

———. 1989, ApJ, 336, 808

———. 1990, in IAU Symp. 140, Galactic and Intergalactic Magnetic Fields, ed. R. Beck, R. Wielebinski, & P. P. Kronberg (Dordrecht: Kluwer), 35

———. 1996, ApJ, 466, 224

———. 2001, PASP, 113, 1243

———. 2002, in ASP Conf. Ser. 278, Single-Dish Radio Astronomy: Techniques and Applications, ed. S. Stanimirović, D. Altschuler, P. Goldsmith, & C. Salter (San Francisco: ASP), 131

———. 2007, PASP, 119, 643

Heiles, C., & Crutcher, R. 2005, in Cosmic Magnetic Fields, ed. R. Wielebinski & R. Beck (Berlin: Springer), 137

Heiles, C., Goodman, A. A., McKee, C. F., & Zweibel, E. G. 1993, in Protostars and Planets III, ed. E. H. Levy & J. I. Lunine (Tucson: Univ. Arizona Press), 279

Heiles, C., & Stevens, M. 1986, ApJ, 301, 331

Heiles, C., & Troland, T. H. 1982, ApJ, 260, L23

———. 2004, ApJS, 151, 271

———. 2005, ApJ, 624, 773

Heiles, C., et al. 2001a, PASP, 113, 1247

———. 2001b, PASP, 113, 1274

Heitsch, F., Zweibel, E. G., Mac Low, M.-M., Li, P., & Norman, M. L. 2001, ApJ, 561, 800

Heyer, M., Gong, H., Ostriker, E., & Brunt, C. 2008, ApJ, in press (astro-ph/0802.2084)

Huguenin, G. R., Taylor, J. H., & Troland, T. H. 1970, ApJ, 162, 727

IAU 1974, in Transactions of the IAU, Vol. XVB 1973, Proceedings of the Fifteenth General Assembly, ed. G. Contopoulos & A. Jappel (Dordrecht: Reidel), 165

IEEE 1997, IEEE Standard Definitions of Terms for Radio Wave Propagation, IEEE Std. 211-1997

Jackson, J. D. 1998, Classical Electrodynamics (3rd ed.; New York: Wiley)

Jenkins, F. A., & Segrè, E. 1939, Physical Review, 55, 52

Jones, R. C. 1941, J. Opt. Soc. Am., 31, 488

———. 1947, J. Opt. Soc. Am., 37, 107

Kalberla, P. M. W., Burton, W. B., Hartmann, D., Arnal, E. M., Bajaja, E., Morras, R., & Pöppel, W. G. L. 2005, A&A, 440, 775

Kazes, I., & Crutcher, R. M. 1986, A&A, 164, 328

Kazes, I., Troland, T. H., & Crutcher, R. M. 1991, A&A, 245, L17

Kazes, I., Troland, T. H., Crutcher, R. M., & Heiles, C. 1988, ApJ, 335, 263

Kemp, J. C. 1970, ApJ, 162, L69

Kennicutt, R. C., Jr. 1998, ARA&A, 36, 189

Killeen, N. E. B., Staveley-Smith, L., Wilson, W. E., & Sault, R. J. 1996, MNRAS, 280, 1143

Kliger, D. S., Lewis, J. W., & Randall, C. E. 1990, Polarized Light in Optics and Spectroscopy (Boston: Academic Press)

Koopmans, L. V. E., Treu, T., Bolton, A. S., Burles, S., & Moustakas, L. A. 2006, ApJ, 649, 599

Kox, A. J. 1997, Eur. J. Phys., 18, 139

Kraus, J. D. 1966, Radio Astronomy (1st ed.; New York: McGraw-Hill)

Landé, A. 1921, Z. Phys., 5, 231

Landi degl'Innocenti, E., & Landolfi, M. 2004, Polarization in Spectral Lines (Dordrecht: Kluwer)

Le Brun, V., Bergeron, J., Boisse, P., & Deharveng, J. M. 1997, A&A, 321, 733

Leroy, J.-L. 2000, Polarization of Light and Astronomical Observation (Amsterdam: Gordon & Breach)

Lo, K. Y. 2005, ARA&A, 43, 625

Lockett, P., & Elitzur, M. 2008, ApJ, 677, 985

Lonsdale, C. J., Diamond, P. J., Thrall, H., Smith, H. E., & Lonsdale, C. J. 2006, ApJ, 647, 185

Lonsdale, C. J., Lonsdale, C. J., Diamond, P. J., & Smith, H. E. 1998, ApJ, 493, L13

Lotz, J. M., et al. 2008, ApJ, 672, 177

Marion, J. B. 1965, Classical Electromagnetic Radiation (New York: Academic Press)

McKee, C. F., & Ostriker, E. C. 2007, ARA&A, 45, 565

Meiring, J. D., et al. 2006, MNRAS, 370, 43

Mirabel, I. F., & Sanders, D. B. 1987, ApJ, 322, 688

Modjaz, M., Moran, J. M., Kondratko, P. T., & Greenhill, L. J. 2005, ApJ, 626, 104

Morganti, R., et al. 2006, MNRAS, 371, 157

Morris, D., Clark, B. G., & Wilson, R. W. 1963, ApJ, 138, 889

Morris, D., Radhakrishnan, V., & Seielstad, G. A. 1964, ApJ, 139, 551

Mouschovias, T. C. 1976, ApJ, 207, 141

Mueller, H. 1948, J. Opt. Soc. Am., 38, 661

Myers, P. C., & Goodman, A. A. 1991, ApJ, 373, 509

Nafe, J. E., & Nelson, E. B. 1948, Phys. Rev., 73, 718

Nakamura, F., & Li, Z.-Y. 2005, ApJ, 631, 411

Napier, P. J. 1999, in ASP Conf. Ser. 180, Synthesis Imaging in Radio Astronomy II, ed. G. B. Taylor, C. L. Carilli, & R. A. Perley (San Francisco: ASP), 37

Narayanan, G., Heyer, M. H., Brunt, C., Goldsmith, P. F., Snell, R., & Li, D. 2008, ApJ, in press (arXiv:0802.2556)

Ostriker, E. C., Stone, J. M., & Gammie, C. F. 2001, ApJ, 546, 980

Padoan, P., Goodman, A., Draine, B. T., Juvela, M., Nordlund, Å., & Rögnvaldsson, Ö. E. 2001, ApJ, 559, 1005

Pais, A. 1986, Inward Bound: Of Matter and Forces in the Physical World (Oxford: Clarendon Press)

Parker, E. N. 1970, ApJ, 160, 383

Parra, R., Conway, J. E., Diamond, P. J., Thrall, H., Lonsdale, C. J., Lonsdale, C. J., & Smith, H. E. 2007, ApJ, 659, 314

Parra, R., Conway, J. E., Elitzur, M., & Pihlström, Y. M. 2005, A&A, 443, 383

Paschen, F., & Back, E. 1912, Ann. d. Phys., 39, 897

Pauli, W. 1925, Z. Phys., 31, 373

Pawsey, J. L., & Bracewell, R. N. 1955, Radio Astronomy (Oxford: Clarendon Press)

Perrin, F. 1942, J. Chem. Phys., 10, 415

Piddington, J. H. 1961, Radio Astronomy (New York: Harper)

Pihlström, Y. M., Baan, W. A., Darling, J., & Klöckner, H.-R. 2005, ApJ, 618, 705

Pihlström, Y. M., Conway, J. E., Booth, R. S., Diamond, P. J., & Polatidis, A. G. 2001, A&A, 377, 413

Press, W. H., Teukolsky, S. A., Vetterling, W. T., & Flannery, B. P. 1992, Numerical Recipes in FORTRAN: The Art of Scientific Computing (2nd ed.; Cambridge: Cambridge Univ. Press)

Preston, T. 1898, Phil. Mag., 45, 333

Price, D. J., & Bate, M. R. 2008, MNRAS, 385, 1820

Raimond, E., & Eliasson, B. 1969, ApJ, 155, 817

Randell, J., Field, D., Jones, K. N., Yates, J. A., & Gray, M. D. 1995, A&A, 300, 659

Reid, M. J., & Moran, J. M. 1988, in Galactic and Extragalactic Radio Astronomy, ed. G. L. Verschuur & K. I. Kellermann (2nd ed.; Berlin: Springer), 255

Reid, M. J., & Silverstein, E. M. 1990, ApJ, 361, 483

Roberts, M. S., Brown, R. L., Brundage, W. D., Rots, A. H., Haynes, M. P., & Wolfe, A. M. 1976, AJ, 81, 293

Robishaw, T., & Heiles, C. 2006, GBT Technical Memo 244

Rogers, A. E. E., Moran, J. M., Crowther, P. P., Burke, B. F., Meeks, M. L., Ball, J. A., & Hyde, G. M. 1967, ApJ, 147, 369

Rohlfs, K., & Wilson, T. L. 1996, Tools of Radio Astronomy (2nd ed.; Berlin: Springer)

Rovilos, E., Diamond, P. J., Lonsdale, C. J., Lonsdale, C. J., & Smith, H. E. 2003, MNRAS, 342, 373

Rudge, A. W., & Adatia, N. A. 1978, Proc. IEEE, 66, 1592

Rybicki, G. B., & Lightman, A. P. 1979, Radiative Processes in Astrophysics (New York: Wiley)

Sarma, A. P., Momjian, E., Troland, T. H., & Crutcher, R. M. 2005, AJ, 130, 2566

Sault, R. J., Killeen, N. E. B., Zmuidzinas, J., & Loushin, R. 1990, ApJS, 74, 437

Savage, B. D., & Sembach, K. R. 1996, ARA&A, 34, 279

Schwarz, U. J., Troland, T. H., Albinson, J. S., Bregman, J. D., Goss, W. M., & Heiles, C. 1986, ApJ, 301, 320

Shinnaga, H., & Yamamoto, S. 2000, ApJ, 544, 330

Shu, F. 1991, The Physics of Astrophysics, Vol. I: Radiation (New York: Univ. Science Books)

Shu, F. H., Allen, R. J., Lizano, S., & Galli, D. 2007, ApJ, 662, L75

Slysh, V. I., & Migenes, V. 2006, MNRAS, 369, 1497

Smith, H. E., Lonsdale, C. J., Lonsdale, C. J., & Diamond, P. J. 1998, ApJ, 493, L17

Soleillet, P. 1929, Ann. Phys., 12, 23

Sommerfeld, A. 1954, Optics: Lectures on Theortical Physics, Vol. IV (New York: Academic Press)

Spitzer, L. 1978, Physical Processes in the Interstellar Medium (New York: Wiley)

Staveley-Smith, L., Cohen, R. J., Chapman, J. M., Pointon, L., & Unger, S. W. 1987, MNRAS, 226, 689

Staveley-Smith, L., Norris, R. P., Chapman, J. M., Allen, D. A., Whiteoak, J. B., & Roy, A. L. 1992, MNRAS, 258, 725

Steidel, C. C., Pettini, M., Dickinson, M., & Persson, S. E. 1994, AJ, 108, 2046

Steinberg, J.-L., & Lequeux, J. 1963, Radio Astronomy (New York: McGraw-Hill)

Stokes, G. G. 1852, Trans. Cambridge Phil. Soc., 9, 399

Stutzman, W. L. 1993, Polarization in Electromagnetic Systems (Boston: Artech House)

Sun, X. H., Reich, W., Waelkens, A., & Enßlin, T. A. 2008, A&A, 477, 573

Thompson, A. R., Moran, J. M., & Swenson, G. W., Jr. 2001, Interferometry and Synthesis in Radio Astronomy (2nd ed.; New York: Wiley)

Thompson, T. A., Quataert, E., & Waxman, E. 2007, ApJ, 654, 219

Thompson, T. A., Quataert, E., Waxman, E., Murray, N., & Martin, C. L. 2006, ApJ, 645, 186

Thomson, J. J. 1897, Phil. Mag., 44, 293

Tinbergen, J. 1996, Astronomical Polarimetry (Cambridge: Cambridge Univ. Press)

———. 2003, Ap&SS, 288, 3

Townes, C. H., & Schawlow, A. L. 1975, Microwave Spectroscopy (New York: Dover)

Troland, T. H. 1990, in IAU Symp. 140, Galactic and Intergalactic Magnetic Fields, ed. R. Beck, R. Wielebinski, & P. P. Kronberg (Dordrecht: Kluwer), 293

Troland, T. H., & Crutcher, R. M. 2008, ApJ, in press (arXiv:0802.2253)

Troland, T. H., Crutcher, R. M., Goodman, A. A., Heiles, C., Kazes, I., & Myers, P. C. 1996, ApJ, 471, 302

Troland, T. H., Crutcher, R. M., & Kazes, I. 1986, ApJ, 304, L57

Troland, T. H., & Heiles, C. 1982a, ApJ, 252, 179

———. 1982b, ApJ, 260, L19

Tufte, S. L., Reynolds, R. J., & Haffner, L. M. 1998, ApJ, 504, 773

Turner, B. E., & Heiles, C. 2006, ApJS, 162, 388

Uchida, K. I., Fiebig, D., & Güsten, R. 2001, A&A, 371, 274

Uhlenbeck, G. E., & Goudsmit, S. 1925, Naturw., 13, 953

van de Hulst, H. C. 1957, Light Scattering by Small Particles (New York: Wiley)

van de Kamp, P. 1967, Principles of Astrometry (San Francisco: W. H. Freeman)

Veilleux, S., Kim, D.-C., & Sanders, D. B. 1999, ApJ, 522, 113

Verschuur, G. L. 1968, Phys. Rev. Lett., 21, 775

———. 1969a, ApJ, 156, 861

———. 1969b, Nature, 223, 140

———. 1979, Fund. Cosmic Phys., 5, 113

———. 1995a, ApJ, 451, 624

———. 1995b, ApJ, 451, 645

Vlemmings, W. H. T. 2007, in IAU Symp. 242, Astrophysical Masers and their Environments, ed. J. M. Chapman & W. A. Baan (Cambridge: Cambridge Univ. Press), 37

———. 2008, A&A, in press (arXiv:0804.1141)

Walker, J. 1904, The Analytical Theory of Light (Cambridge: Cambridge Univ. Press)

Weiler, K. W. 1973, A&A, 26, 403

White, H. E. 1934, Introduction to Atomic Spectra (New York: Wiley)

Wolfe, A. M., Broderick, J. J., Condon, J. J., & Johnston, K. J. 1976, ApJ, 208, L47

Wolfe, A. M., Gawiser, E., & Prochaska, J. X. 2005, ARA&A, 43, 861

Wolfe, A. M., Prochaska, J. X., & Gawiser, E. 2003, ApJ, 593, 215

Wolleben, M., & Reich, W. 2004a, in The Magnetized Interstellar Medium, ed. B. Uyaniker, W. Reich, & R. Wielebinski (Katlenburg-Lindau: Copernicus GmbH), 99

———. 2004b, A&A, 427, 537

Yu, Z.-Y. 2004, Chinese Astron. Astrophys., 28, 287

———. 2005, Chinese J. Astron. Astrophys., 5, 159

Zeeman, P. 1897a, Phil. Mag., 43, 226

———. 1897b, ApJ, 5, 332

———. 1897c, Phil. Mag., 44, 55

———. 1931, Nature, 128, 365

Zweibel, E. G. 1990, ApJ, 362, 545

Index

www.ingramcontent.com/pod-product-compliance
Lightning Source LLC
Chambersburg PA
CBHW080552220326

41599CB00032B/6455